Problematic Soils and Their Management

a division of
NIPA GENX ELECTRONIC RESOURCES & SOLUTIONS P. LTD.
New Delhi-110 034

About the Editors

Abhay Kumar is currently working as an Assistant Professor (Agroforestry) in Department of Agronomy, Ranchi Agriculture College, Birsa Agricultural University, Ranchi, Jharkhand. He has completed B.Sc. (Hons) Forestry from NAU, Gujarat and M.Sc. Agroforestry from GBPUAT, Pantnagar, Uttarakhand and Ph.D. Forestry (Pursuing), specialization in Agroforestry, BAU, Ranchi. He has also qualified NET three times from ASRB-ICAR, New Delhi and UGC-NET in Environmental science. He has been awarded with Indo Global Excellence Award, Young Teacher Award and Young Scientist Award-2018 and also got NTS from July 2009 to June 2013 by ICAR. He has participated in and completed more than 10 online MOOC courses conducted by IIRS, NAARM, and Elsevier etc. He is life member of many societies and editorial member of journals. He has got more than 15 research papers, 2 books, 3 book chapters and a number of articles in magazine/bulletin published and more than 20 seminar/symposia presented or attended.

Dr. (Mrs.) Asha Kumari Sinha is currently working as an Assistant Professor cum-Junior Scientist in the Department of Soil Science and Agricultural Chemistry, Ranchi Agriculture College, Birsa Agricultural University, Ranchi, Jharkhand. She has completed B.Sc.(Hons) Agriculture, M.Sc. and Ph.D. in Soil Science from Birsa Agricultural University, Ranchi, Jharkhand. She has been awarded with Outstanding Achievement Award-2018. She has participated and completed three trainings (21 days duration) conducted by ICAR and more than five short duration trainings. She has also participated and completed online MOOC course. She is a life member of Indian Society of Soil Science. She has got more than 20 research papers and a number of articles in magazine/bulletins and more than 20 seminar/symposia presented or attended. She has guided more than six M.Sc. students under their research work as a Major Advisor/Co-Advisor/Minor Advisor.

Swati Shabnam is currently working as an Assistant Professor (Agronomy) in Department of Agronomy, Ranchi Agriculture College, Birsa Agricultural University, Ranchi, Jharkhand. She has completed B.Sc. (Hons) Agriculture from SHUATS, Prayagraj, U.P., and M.Sc. and Ph.D. Agronomy from Birsa Agricultural University, Ranchi. She has qualified NET from ASRB-ICAR, New Delhi. She has published many research papers, 2 books, 2 book chapters, magazine/bulletin and participated/presented paper in many national/international seminar and symposia.

Problematic Soils and Their Management

Editors

Abhay Kumar
Assistant Professor
Department of Agronomy
Ranchi Agriculture College
Birsa Agricultural University, Kanke
Ranchi, Jharkhand-834006, India

Asha Kumari Sinha
Assistant Professor (Soil Science)
Department of Soil Science & Agricultural Chemistry
Ranchi Agriculture College
Birsa Agricultural University
Kanke, Ranchi, Jharkhand-834006, India

Swati Shabnam
Assistant Professor (Agronomy)
Department of Agronomy
Ranchi Agriculture College
Birsa Agricultural University, Kanke
Ranchi, Jharkhand-834006, India

a division of

NIPA GENX ELECTRONIC RESOURCES & SOLUTIONS P. LTD.
New Delhi-110 034

a division of
NIPA GENX ELECTRONIC RESOURCES & SOLUTIONS P. LTD.
101,103, Vikas Surya Plaza, CU Block
L.S.C.Market, Pitam Pura, New Delhi-110 034
Ph : +91 11 27341616, 27341717, 27341718
E-mail:newindiapublishingagency@gmail.com
www: www.nipabooks.com

For customer assistance, please contact
Phone: + 91-11-27 34 17 17 Fax: + 91-11- 27 34 16 16
E-Mail: feedbacks@nipabooks.com

ISBN: 978-93-90591-38-1

Composed and Designed by NIPA.

Birsa Agricultural University
Kanke, Ranchi-834006, Jharkhand, India

Dr. Onkar Nath Singh
Vice Chancellor

Foreword

Importance of soil health is well established in all dimensions. 68th UN General Assembly declared 2015 as the International Year of Soils. FAO has formed Global soil partnership with various countries to promote healthy soils for a healthy life and world without hunger. India, the second most populous country in the world faces severe problems in agriculture like deteriorating soil health and decreasing production. In February 2015, Government of India launched the scheme of Soil Health Card which is used to assess the current status of soil health, i.e. nutrient status and soil pH of farmer's filed which helps farmers use judicious amount of fertilizers which in turn will be helpful to increase farmers' income, reduce environmental pollution by excess use of fertilizers and when used over time, it can determine changes in soil health affected by land management practices.

Problematic soils are soils which possess characteristics, like acidity, salinity-alkalinity, high calcium content, waterlogged condition, soil degradation, pollution etc., that make them uneconomical for crop cultivation. Soils of large area (around 67.45 m ha) are considered to be problematic and proper management and reclamation of these soils is necessary for cultivation of crops as well as to prevent irreparable damage to the soil health. Healthy soils provide a range of environmental services including water infiltration, habitat provision, and profitable and sustainable agriculture.

I am therefore happy that the present book on **"Problematic Soils and Their Management"** compiled or written by Abhay Kumar, Asha Kumari Sinha and Swati Shabnam has brought together a summary of the works done in this field. We owe a deep debt of gratitude to the authors for their labour of love for the science.

We firmly believe that this publication will be highly useful, in one way or other, for researchers, academicians, extension workers, policy makers, planners, officials in development institutions/agencies, producers, farmers and especial for UG B.Sc. (Agri) students of agriculture.

O. N. Singh

Acknowledgement

The authors express their sincere gratitude to Dr. Onkar Nath Singh, Hon'ble Vice Chancellor, Birsa Agricultural University, Ranchi, Dr. M. S. Yadava Dean, Faculty of Agriculture, Dr. M. H. Siddaqui, Dean, Faculty of Forestry Dr. M. S. Malik, University Professor and Chairman, Faculty of Forestry Dr. P. R. Oraon, Assistant Registrar, Faculty of Forestry, Birsa Agricultural University, Ranchi, Jharkhand. Dr Virendra Singh, University Professor Dr. Salil Tiwari, Dean Student Welfare, GBPUAT, Pantnagar, Uttarakhand Dr. Suman Jha, Dr. Satish Sinha, Dr. N. S. Thakur, College of Forestry, Navsari Agricultural University, Gujarat and Dr. Ashok Kumar, PAU, Ludhiana Panjab. We also thank our authors, Dr. Rakesh Kumar, Dr. R.R. Upasani Dr. Sheela Barla, Mr. Asisan Minz, Mr. Deo Kumar, Ms. Shikha Verma, Ms. Reshma Shinde, Mr. Pradeep Kr. Sarkar, Dr. S. Firdous, Dr. Vikas Rena Dr. Shailendra Kumar, Dr. Ram Gopal, Mr. Indra Singh, Ms. Sunita Kumari D. R. Prajapati, Ms. Parthasarathi T., Mr. Bijay Kumar Singh for their valuable contribution of chapters. The editors also acknowledge all persons involved directly or indirectly in preparation of this work.

Preface

This book "Problematic Soils and Their Management" in its small volume deals with the problems of students and faculty members of soil science to meet the immediate needs of fourth semester B. Sc. (Agri) programme as per the syllabus prescribed by the Fifth Deans' committee on Higher Agricultural Education in India, effective from the academic years 2016-17.

The soils which possess characteristics that make them uneconomical for the cultivation of crops without adopting proper reclamation measures are known as problem soils. For the management of problematic soils, some general principles have to be considered for proper implementation of the reclamation measures. Some soils have serious physical and chemical limitations to cultivation. Physical limitations can be managed by irrigation, drainage mulching, manuring, tillage, and soil conservation measures such as terracing contouring, and cover crops, whichever is appropriate. Some chemicals are also added to the soil as an integral part of the reclamation program adopted to improve the saline, alkali and other soils.

This book contains 19 chapters on soil quality and health, Distribution of Waste land and problem soils in India and Jharkhand, Soil forming factors and Processes Reclamation and management of Saline and sodic soils, Acid soils and Acid Sulphate soils, Calcareous soils, Eroded and Compacted soils, Flooded soils, Polluted soils, Irrigation water-quality and standards, Efficient utilization of saline water in agriculture, Remote sensing and GIS in diagnosis and management of problem soils, Multipurpose tree species, Bio remediation through MPTs of soils, land capability land suitability classification, Problematic soils under different Agro-ecosystems, etc

Keeping in mind these points, we wrote the book on "Problematic Soils and Their Management" and suggest UG students of fourth semester B.Sc. (Agri) as per the syllabus prescribed by the Fifth Deans committee.

Editors

Contents

List of Symbols Used

Notations used	Description	Notations used	Description
@	At the rate of	ml	Millilitre
a.i	Active ingredient	MT	Million tons
b	Billion	MOP	Muriate of potash
cm	Centimeter	mV	Milli Volts
dS/m	Deci Siemens per metre	N	Nitrogen
et al.	Coworkers and others	NPK	Nitrogen, phosphorus and potassium
^{0}C	Degree Celsius		
EC	Electrical conductivity	No.	Number
ESP	Exchangeable sodium percentage	OC	Organic carbon
		pH	Negative logarithm of the reciprocal of the H^+ ion activity
etc.	Et. cetera		
f.b.	Followed by	%	Percent
Fig.	Figure	P_2O_5	Phosphorus
FYM	Farm Yard Manure	ppm	Parts per million
g	Gram	K_2O	Potash
ha	Hectare	Rs.	Rupees
h	Hour	S	Second
Kcal	Kilo calorie	SSP	Single super phosphate
kg	Kilogram	m^2	Square metre
km	Kilometre	Temp.	Temperature
L	Litre	Viz.	That is to say/ in others words
m	Metre	t	Tons
m ha	Million Hectare	i.e.	That is
mm	Millimetre	Yr	Year

List of Contributors

Abhay Kumar, *Assistant Professor (Agroforestry) Birsa Agricultural University, Kanke, Ranchi Jharkhand-834006, India*

Asha Kumari Sinha, *Assistant Professor cum Junior Scientist (Soil Science), Birsa Agricultural University, Kanke, Ranchi, Jharkhand-834006, India*

Asisan Minz, *Assistant Professor cum Junior Scientist (Soil Science), RNTAC, Deoghar, Birsa Agricultural University, Kanke, Ranchi, Jharkhand-834006, India*

Bijay Kumar Singh, *Senior Research Fellow AICRP (Agroforestry), Birsa Agricultural University Kanke, Ranchi, Jharkhand-834006, India*

D. R. Prajapati, Ph.D. Scholar *(Agroforestry), Navsari Agricultural University, Navsari-396450, Gujarat, India*

Deo Kumar, *Assistant Professor cum Junior Scientist (Soil Science), Birsa Agricultural University Kanke, Ranchi, Jharkhand-834006, India*

Kerobim Lakra, *Assistant Professor (Agri. Economics), Birsa Agricultural University, Kanke Ranchi, Jharkhand-834006, India*

M. S. Yadava, *University Professor & Dean (Agriculture), Birsa Agricultural University, Kanke Ranchi, Jharkhand-834006, India*

M.H. Siddiqui, *University Professor & Dean (Forestry), Birsa Agricultural University, Kanke Ranchi, Jharkhand-834006, India*

M.S. Malik, *University Professor & Chairman (Agroforestry), Birsa Agricultural University, Kanke Ranchi, Jharkhand-834006, India*

Parthasarathi T., *Assistant Professor (Soil Science), Vellore Institute of Technology, Vellore-632014 Tamil Nadu, India*

Prabhat Ranjan Oraon, *Assistant Professor cum-Junior Scientist (Forestry), Birsa Agricultural University, Kanke, Ranchi, Jharkhand-834006, India*

Pradip Kumar Sarkar, *Scientist (Agroforestry), ICAR- Research Complex for Eastern Region FSRCHPR, Ranchi-834010, Jharkhand, India*

R. R. Upasani, *University Professor (Agronomy), Birsa Agricultural University, Kanke, Ranchi Jharkhand-834006, India*

Rakesh Kumar, *University Professor (Soil Science), Birsa Agricultural University, Kanke, Ranchi Jharkhand-834006, India*

Ram Gopal, Ph.D. Scholar *(Forestry), Forest Research Institute, Dehradun, P.O. New Forest-248006 Uttarakhand, India*

Reshma Shinde, *Scientist (Soil Science), ICAR-Research Complex for Eastern Region, FSRCHPR Ranchi-834010, Jharkhand, India*

Saiyyeda Firdous, *Assistant Professor (Soil Science), Vellore Institute of Technology, Vellore-632014 Tamil Nadu, India*

Salil Kumar Tewari, *University Professor and Dean Student Welfare (Agroforestry), G. B. Pant University of Agriculture & Technology, Pantnagar, Uttarakhand-263145, India*

Shailendra Kumar, Ph.D. Scholar *(Environmental Science), Jawaharlal Nehru University, New Delhi-110067, India*

Sheela Barla, *Assistant Professor cum Junior Scientist (Agronomy), Birsa Agricultural University Kanke, Ranchi, Jharkhand-834006, India*

Shikha Verma, Ph.D. Scholar *(Soil Science), Birsa Agricultural University, Kanke, Ranchi Jharkhand-834006, India*

Shushama Majhi, *Assistant Professor cum Junior Scientist (Agronomy), Birsa Agricultural University, Kanke, Ranchi, Jharkhand-834006, India*

Soumitra Sankar Das, *Assistant Professor cum Junior Scientist (Agri. Statistics), Birsa Agricultural University, Kanke, Ranchi, Jharkhand-834006, India*

Sunita Kumari, Ph.D. Scholar *(Forestry), Institute of Forest Productivity, Ranchi, Jharkhand- 835303, India*

Swati Shabnam, *Assistant Professor (Agronomy), Birsa Agricultural University, Kanke, Ranchi Jharkhand-834006, India*

Vikas Rena, Ph.D. Scholar *(Environmental Science), Jawaharlal Nehru University, New Delhi-110067, India*

Virendra Singh, *University Professor (Agroforestry), G. B. Pant University of Agriculture & Technology, Pantnagar, Uttarakhand-263145, India*

1

Soil Quality and Health

Abhay Kumar, M.S. Malik, M.H. Siddiqui and M.S. Yadava

Introduction

Intensive production of agricultural crops has contributed to decline in soil quality, leading to lower crop productivity and farm profitability. Major causes of this decline are soil compaction, surface crusting, low organic matter, increased pressure and damage from diseases, weeds, insects, and other pests, as well as a lower density and diversity of beneficial soil organisms. These constraints have increased the interest of farmers in assessing the health status of their soils and in implementing sustainable soil management practices. Soil health is critically important to sustainable agricultural productivity and environmental well being. Healthy soils provide a range of environmental services including water infiltration, habitat provision and profitable and sustainable agriculture.

Soil quality

Soil quality is the capacity of a specific kind of soil to function, within natural or managed ecosystem boundaries, to sustain plant and animal productivity, maintain or enhance water and air quality, and support human health and habitation (Karlen *et al.,*1997). Dynamic soil quality, on the other hand, generally refers to the condition of soil that is changeable in a short period of time by human impact, including agricultural management practices (Karlen *et al.,* 2003). Soil quality can be viewed in two different ways: as an inherent attribute of soils that can be inferred from soil characteristics or indirect observations; or as a capacity to perform certain productivity, environmental, and health functions. Intrinsic soil properties governing soil resilience are related to soil quality. There has been considerable national and international interest in soil quality and health as a key issue relating to agricultural sustainability. Direct effects of soil quality on water quality are attributed to inherent soil characteristics, e.g., parent material, texture, and structure. Land use and soil management also affect water quality through the effects of soil quality. The magnitude of the effect of soil quality on water quality is greatly modified by management systems.

Soil health

Soil health is the capacity of soil to function as a vital living system within ecosystem and land-use boundaries, to sustain plant and animal productivity, maintain or enhance water and air quality, and promote plant and animal health (Doran and Zeiss, 2000). To evaluate sustainability of agricultural practices, assessment of soil health using various indicators of soil quality is needed. Soil organism and biotic parameters (e.g. abundance, diversity, food web structure, or community stability) meet most of the five criteria for useful indicators of soil quality. Several farmer participatory programs for managing soil quality and health have incorporated abiotic and simple biotic indicators. The challenge for the future is to develop sustainable management systems which are the vanguard of soil health; soil quality indicators.

With respect to farmer, the term "soil health" is often preferred over soil quality as it connotes a holistic approach to soil management, including the integration of physical, biological and chemical processes (Idowu *et al.*, 2007). In the past, an overemphasis on chemical soil management has resulted in a loss of the biological and physical fertility of the soil. A new emphasis on soil health through the linkages between the chemical, biological and physical processes (Fig -1.1) therefore provides a more useful framework for sustainable soil management for diverse cropping systems with tools such as organic and inorganic fertilizers, reduced tillage, cover cropping, new rotations, etc.

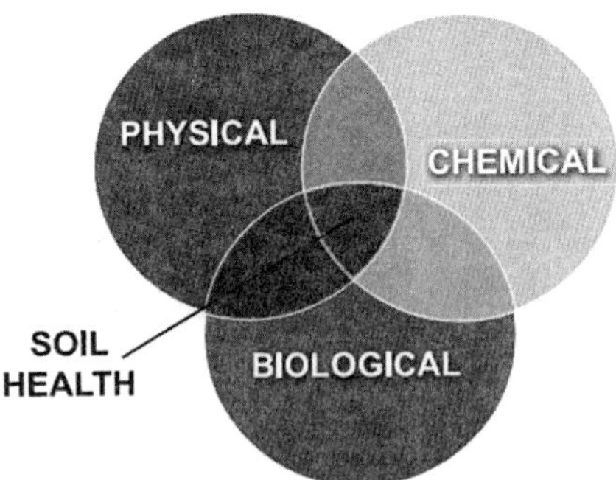

Fig. 1.1: The concept of soil health deals with integrating the physical, biological and chemical components of the soil. Adapted from the Rodale Institute

Soil quality and health

The terms 'soil health' and 'soil quality' are becoming increasingly familiar worldwide. Soil health is defined as "the continued capacity of the soil to function as a vital living ecosystem that sustains plants, animals and humans" and soil quality as "the capacity of a soil to function, within ecosystem and land use boundaries, to sustain productivity, maintain environmental quality and promote plant and animal health."

In general, soil health and soil quality are considered synonymous and can be used interchangeably, with one key distinction conceptualized by scientists and practitioners over the last decades. The term 'soil health' has been generally preferred by farmers, while scientists have generally preferred 'soil quality'. Soil quality includes both inherent and dynamic quality. Inherent soil quality refers to the aspects of soil quality relating to a soil's natural composition and properties, influenced by the natural long term factors and processes of soil formation. Dynamic soil quality, which is equivalent to soil health, refers to soil properties that change as a result of soil use and management over the human time scale. Soil health invokes the idea that soil is an ecosystem full of life that needs to be carefully managed to regain and maintain our soil's ability to function optimally.

Characteristics of soil quality and healthy soil

- Good soil tilth
- Sufficient depth
- Good water storage and good drainage
- Sufficient supply, but not excess of nutrients
- Small population of plant pathogens and insect pests
- Large population of beneficial organisms
- Low weed pressure
- Free of chemicals and toxins that may harm the crop
- Resistant to degradation
- Resilience when unfavorable conditions occur

Following factors affect of soil quality and health

- Soil Compaction
- Poor Aggregation
- Weed Pressure
- High Pathogen Pressure
- Low Water and Nutrient Retention

- Salinity and Sodicity
- Heavy Metal Contamination
- Industrial waste

Soil health indicators

Soil health is best assessed through soil properties that are sensitive to changes in management. Thus, soil health assessment evaluates the soil's ability to accommodate most of the relevant processes relevant to crop production and soil hydrology. The Cornell soil health program of the Cornell University, New York, has done significant work on soil health assessment. The program evaluated 39 potential soil health indicators (Table-1.1) and selected 12 measures for inclusion in the soil health assessment program. These measures or properties can be considered as indicators of different soil processes (Table-1.2).

The selection of these measures is based on

- sensitivity to changes in soil management practices
- relevance to soil processes and functions
- consistency and reproducibility
- ease and cost of sampling
- cost of analysis

Table 1.1: Thirty nine potential indicators evaluated for use in the Cornell soil health assessment protocol (Gugino *et al.*, 2009)

Indicator	Indicator
Physical Indicators	Parasitic nematode population
Bulk density	Potential mineralizable nitrogen
Macro-porosity	Decomposition rate
Meso-porosity	Particulate organic matter
Micro-porosity	Active carbon test
Available water capacity	Weed seed bank
Residual porosity	Microbial respiration rate
Penetration resistance at 10 kPa	Glomalin content
Saturated hydraulic conductivity	Chemical Indicators
Dry aggregate size (<0.25 mm)	pH
Dry aggregate size (0.25 - 2 mm)	Phosphorus
Dry aggregate size (2 - 8 mm)	Nitrate Nitrogen
Wet aggregate stability (0.25 -2 mm)	Potassium
Wet aggregate stability (2 - 8 mm)	Magnesium
Surface hardness (penetrometer)	Calcium

Indicator	Indicator
Subsurface hardness (penetrometer)	Iron
Field infiltrability	Aluminum
Biological Indicators	Manganese
Root health assessment	Zinc
Organic matter content	Copper
Beneficial nematode population	Exchangeable acidity

Table 1.2: Indicators of physical, biological and chemical health of soil and their respective soil processes (Idowu *et al.*, 2007)

Soil health assessment indicator	Soil functional processes
Physical Indicators	
Aggregate Stability	Aeration, infiltration, shallow rooting, crusting
Available Water Capacity	Water retention
Surface Hardness	Rooting, water transmission
Subsurface Hardness	Rooting at depth
Biological Indicators	
Organic Matter Content	Energy / C storage, water and nutrient retention
Active Carbon Content	Organic material to support biological functions
Potentially Mineralizable Nitrogen (PMN)	N supply capacity, N leaching potential
Root Health Rating	Soil-borne pest pressure
Chemical Indicators	
pH	Toxicity, nutrient availability
Extractable Phosphorus	P availability, environmental loss potential
Extractable Potassium	K availability
Minor Element Contents (4 elements)	Micronutrient availability, element imbalances

Soil health cards

Soil health or soil quality assessment card is a qualitative tool designed for farmers. Soil health cards are farmer friendly, quick, and require only basic tools. Results are obtained immediately, allowing evaluation of numerous fields quickly. Directions for use are found on each card. The cards contain farmer selected soil quality indicators and associated ranking descriptions typical of local farmers. Generally, indicators listed, such as soil tilth, abundance of earthworms, or water infiltration, can be assessed without the aid of technical or laboratory equipment. All cards have a scoring system, which usually includes either a range of poor to good or a numerical scale from 1 to 10 for each indicator.

Health cards integrate physical, biological, and chemical properties in ways that are familiar to farmers. For example, the cards use terms like tilth, which refers to the physical structure of soil and which also depends on biological properties.

In India, Soil Health Card is a scheme launched by the Government of India in 19 February 2015. From 2015 to 2017, in Cycle-1, 10.74 crore soil health cards were distributed to farmers. In Cycle-II (2017-19), 11.69 crore soil health cards have been distributed to farmers across the country. Establishment of Soil Testing Labs: So far 429 New Static Soil Testing Labs (STLs), 102 New Mobile STLs, 8752 Mini STLs have been provided. Soil Health Card is used to assess the current status of soil health and, when used over time, to determine changes in soil health that are affected by land management.

Soil Health Card is a printed report that a farmer will be handed over for each of his holdings. It will contain the status of his soil with respect to 12 parameters, namely N,P,K (Macro-nutrients); S (Secondary-nutrient); Zn, Fe, Cu, Mn, Bo (Micro-nutrients); and pH, EC, OC (Physical parameters). Based on this, the SHC will also indicate fertilizer recommendations and soil amendment required for the farm.

Soil health / Quality kits

The United States Department of Agriculture has developed a soil quality test kit (USDA, 1999). It is a quantitative assessment kit that can provide results to diagnose possible soil problems, such as compaction or salinity, compare management systems and monitor changes in soil quality over time. The kit uses a minimum dataset of indicators chosen primarily for agricultural soils quality assessments, which are integrated into quantitative tests for biological, chemical and physical properties of the soil ecosystem. A total of 11 tests can be performed, including soil respiration, infiltration, bulk density, electrical conductivity, soil pH, soil nitrate content, aggregate stability, soil slaking, earthworm counts, and various observations of soil physical attributes. The kit consists of a portable box, which includes most of the equipment needed to complete the tests. A guide is included in the kit. The kit is used as a screening tool to give a general direction or trend of soil quality. It can also be used to troubleshoot problem areas in the field. Several soil test kits, developed by research institutions and private industry are in use in India.

Soil health report

The results of measurements are synthesized into a farmer friendly soil health report that can initially be used as a baseline assessment. Subsequent sampling and analysis of the same field can be employed to determine the impact of implemented soil management practices on soil health. The sections of the report include:

Background information: Information collected during sampling including the farmer's name and contact information, the sample number, the date of sampling, the local extension educator's name, current crop and tillage and their history over the past 2 years, drainage and slope conditions, soil type and soil texture.

Indicator list: This section gives a list of indicators that were measured for soil health assessment. They are color coded to separate the physical, biological and chemical indicators.

Indicator values: This presents the values of the indicators that were measured either in the laboratory or field.

Ratings: This section presents the scores and color coded ratings of the soil quality indicators. The indicators are scored on a scale of 1-100 based on scoring functions developed for individual indicators. In addition, the indicators are rated with color codes depending on their scores. Generally, a score of less than 30 is regarded as low and receives a red color code. A score from 30 to 70 is considered medium and is color coded yellow. A score value higher than 70 is regarded as high and color coded green.

Constraints: If the rating of a particular indicator is poor/ low (red color code), the respective soil health constraints will be highlighted in this section. This is a very useful tool for identifying areas to target their management efforts.

Overall quality score: An overall quality score is computed as the mean of the individual indicator scores. This score is further rated as follows: less than 40 per cent is regarded as very low, 40-55 per cent is low, 55-70 per cent is medium, 70-85 per cent is high and greater than 85 per cent is regarded as very high. The highest possible quality score is 100 and the least score is 0, thus it is a relative overall soil health status indicator.

Six steps of the soil health management planning process

1. Determine farm background and management history
2. Set goals and sample for soil health
3. For each management unit: identify and explain constraints, prioritize
4. Identify feasible management options
5. Create short and long term Soil Health Management Plan
6. Implement, monitor, and adapt

The management practices for soil health are grouped into 5 major activities called the Soil health management tool box. They are

1. **Tillage considerations**- No Tillage, Ridge Tillage, Strip Tillage, Permanent drive rows, Roller crimpers and rotavators, Cover crop interseeders
2. **Crop rotation considerations**-
   - Grow the same annual crop for only one year
   - Do not grow one crop with another closely related species
   - Use crop sequences that promote healthier crops
   - Follow a legume forage crop(clover or alfalfa), with a high nitrogen-demanding crop
   - Use crop sequences that aid in controlling weeds
   - Grow a deep-rooted crop
   - Grow some crops that will leave a significant amount of residue
3. **Cover cropping considerations**-Cover crops are usually grown for less than one year. They provide a canopy, organic matter inputs, increased species diversity, and living root activity for soil protection and improvement between the productions of main cash crops. Some cover cropping like Winter cover crops, Season-long cover crops, Summer fallow cover crops, Cover crop mixes etc.
4. **Organic considerations-** use of Animal manure, Green manure, Compost, FYM, Vermicompost etc
5. **Considerations for adapting to and mitigating climate change**
   - Soil health management for carbon sequestration: capturing and storing carbon in soils
   - Soil health management to prevent nitrous oxide emissions
   - Soil health management to prevent methane emissions.

Economic benefits of maintaining and improving soil health

- Better plant growth, quality, and yield
- Reduced risk of yield loss during periods of environmental stress (e.g., heavy rain, drought, pest or disease outbreak)
- Better field access during wet periods

- Reduced fuel costs by requiring less tillage
- Reduced input costs by decreasing losses, and improving use efficiency of fertilizer, pesticide, herbicide, and irrigation applications

References

Doran, J. W. and Zeiss, M. R. 2000. Soil health and sustainability: managing the biotic component of soil quality. *Applied Soil Ecology*, 15: 3–11.

Gugino, B. K., Idowu, O. J., Schindelbeck, R. R., van Es, H. M., Wolfe, D. W., Moebius-Clune, B. N., Thies, J. E. and Abawi, G. S. 2009. Cornell Soil Health Assessment Training Manual. Cornell University, New York, USA, pp 57.

Idowu, O. J., van Es, H. M., Schindelbeck, R. R., Abawi, G. S., Wolfe, D. W., Thies, J., Gugino, B. K., Moebius, B. N. and Clune, D. 2007. The new Cornell soil health test: Protocols and interpretation. What's Cropping Up? A news letter for New York field crops and soils, 17(2).

Karlen, D. L., Ditzlerb, C. A. and Andrews, S. S. 2003. Soil quality: why and how? *Geoderma*, 114: 145– 156.

Karlen, D. L., Mausbach, M. J., Doran, J. W., Cline, R. G., Harris, R. F. and Schuman, G. E. 1997. Soil quality: A concept, definition, and framework for evaluation. *Soil Science Society of America Journal,* 61: 4-10.

Moebius-Clune, B. N., Moebius-Clune, D. J., Gugino, B. K., Idowu, O. J., Schindelbeck, R. R., Ristow, A. J., van Es, H. M., Thies, J. E., Shayler, H. A., McBride, M. B., Kurtz, K. S. M., Wolfe, D. W. and Abawi, G. S. 2016. Comprehensive Assessment of Soil Health – The Cornell Framework, Edition 3.2, Cornell University, Geneva, New York, USA.

Srinivas, K. 2018. Techniques of Assessing Soil Health - Physical, Chemical and Biological. In Rao, G. R., Prabhakar, M., Venkatesh, G., Srinivas, I. and Reddy, K. S. Book: Agroforestry Opportunities for Enhancing Resilience to Climate Change in Rainfed Areas. pp. 121-131.

USDA. 1999. Soil quality test kit guide. Soil Quality Institute, National Soil Conservation Service, United States Department of Agriculture, pp. 82.

2

Wasteland and Problem Soils in India

Abhay Kumar, Indra Singh, Ram Gopal and Shailendra Kumar

Introduction

Over 250 million people are directly affected by land degradation worldwide. Each year 12 million hectares are lost to deserts. Only 11 % of the global land surface can be considered as prime or Class-I land.

India contains 2.4 % of world's total land. Out of the total geographical area of 329 million ha of the country, land use statistics are available for about 93 % of the total area which comes around 306 million ha. The arable land (net area sown plus the current and other fallow lands) was estimated at 166.14 million ha (54.2%) while the area under forest was accounted at 69.02 million ha (22.6%). Land put to non-agricultural uses was estimated at 22.97 million ha (7.5%) and the barren and uncultivable land at 19.44 million ha (6.4%). Permanent pastures and other grazing lands where estimated 11.04 million ha. (3.6%), land under miscellaneous uses at 3.62 million ha. (1.2%) and the cultivable waste land at 13.08 million ha (4.5%). By summing up these figures, the reporting area comes out 306 million ha.

It is estimated that only 266 million ha out of the total 329 million ha possess potential of agricultural production. Out of this, 143 million ha is agricultural land while 85 million ha suffers from varying degrees of soil degradation. Of the remaining 123 million ha, 40 million ha are completely unproductive and 83 million ha is classified as forest land. Thus out of 266 million ha, about 175 million ha or 66 % is degraded to varying degrees.

Wasteland

National Wasteland Development Board (NWDB) defined wasteland as degraded land which can be brought under vegetative cover with reasonable effort and which is currently under utilized and land which is deteriorating for lack of appropriate water and soil management or on account of natural causes.

In another definition, wastelands are the degraded and unutilized lands except current fallows due to different constraints. It is estimated that the biomass production in wasteland is less than 20% of its overall potential. It includes area affected by water logging, sheet and gully erosion, riverine lands, shifting cultivation, salinity and alkalinity, wind erosion, shifting and sand dunes, drought etc.

National wastelands development board classifies wastelands into two categories:

i. **Cultivable wasteland-** The land which is capable of, or has potential for the development of vegetative cover but are not being used due to different constraints of varying degrees such as erosion, water logging, salt effects etc.

ii. **Uncultivable wasteland-** The land which cannot be developed for vegetative cover e.g. rocky area, deserts and snow-covered glacial areas.

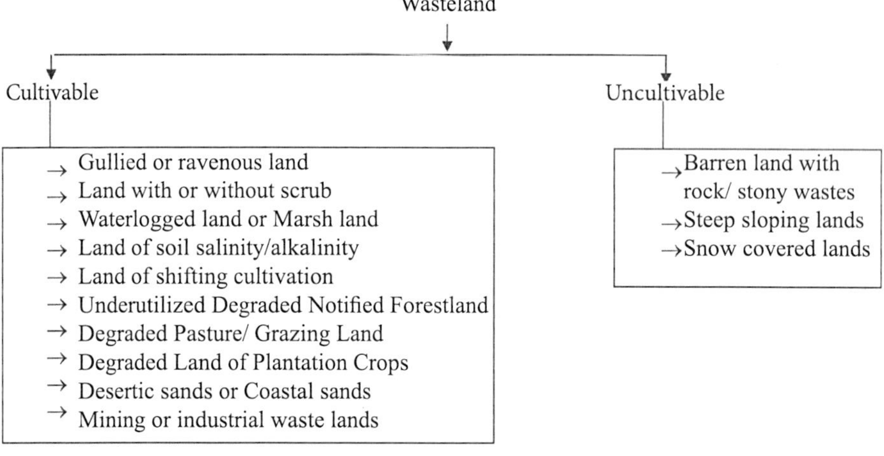

Wasteland mapping techniques

i. Conventional system- The surveyor transverses from field to field with a base map to collect the soil and land use data after closely spaced field observations to identify and map wasteland.

ii. Aerial Photo-interpretation- Black and white panchromatic aerial photographs of 1: 25000 scale and smaller are very suitable for wastelands mapping.

iii. Remote sensing- The satellite era ushers new scopes in land use mapping.

Government of India has set up the National Wastelands Development Board in 1985 under the Ministry of Environment & Forests. Wastelands are lands which are economically unproductive, ecologically unsuitable and subject to environmental deterioration. The official estimate of wasteland in India in 2004 was nearly 63.85 million ha, more than 20% of the country's geographical area.

Table 2.1: Estimated Area under the Wastelands provided by different organization in India

Source	Area (m.ha.)
Ministry of Agriculture and the JNU, Deptt. Of Geography (1986)	175
National Land Use and Wasteland Development Council (First Meeting 1986)	123
Society for Promotion of Wasteland Development (1982)	145
Ministry of Rural Development & NRSA (2000)	64

Table 2.2: The category wise wasteland types of India

Wasteland class	Area (in M ha)	Percentage
Gullied/ravinous land	2.06	0.65
Land with/without scrub	19.40	6.13
Waterlogged/marshy land	1.66	0.52
Land affected by salinity	2.04	0.65
Shifting cultivation area	3.51	1.11
Degraded notified forest land	14.07	4.44
Degraded pastures/grazing land	2.60	0.82
Degraded land under plantation	0.58	0.18
Sandy area	5.00	1.58
Mining/industrial wasteland	0.12	0.04
Barren rock y/stony/sheet rock	6.46	2.04
Steep sloping area	0.77	0.24
Snow covered/glacial area	5.58	1.76
Total	63.85	20.16

Note: total wastelands are estimated at 63.85 M ha correlating with strong, extreme and part of moderate categories of land degradation.
Source: NRSA and MoRD, 2000.

Wasteland assessment

i. Poor fertility of soil due to rocky, sandy, saline, alkali, water logging erosion etc.

ii. Steep and undulated slopes

iii. Shifting cultivation

iv. Frequent droughts

v. Lack of irrigation facilities

vi. Lack of resources

vii. Poor economic conditions

Causes of wasteland

The causes of land degradation has been grouped into three categories-

i. Natural degradation hazards-cyclones, drought, volcanic

ii. Direct causes of land degradation-deforestation, shifting cultivation, overgrazing, non-adoption of soil conservation. Improper crop rotation, unbalanced fertilizer and pesticide

iii. Underlying causes of land degradation- population increase, economic pressure, land tenure, land shortage, poverty

Some other causes of land degradation are follows-

i. Lack of land use planning

ii. Sea level rise

iii. Drought

iv. Unsustainable agricultural practices

v. Uncontrolled waste disposal

Effects of wasteland

i. Decline potential yields

ii. Reduced vegetation cover

iii. Biodiversity loss

iv. Top soil loss

v. Water shortage

vi. Reduce return of organic matter and less biological activity in the soil

vii. Increased pollution from increased use of agrochemicals

viii. Inputs and costs of production are increased

ix. Risk is increased

x. Labour, technical and financial resources are diverted

Reclamation and management of wasteland

1. Land conservation-It is the measures taken to maintain the function of the land.

2. Land rehabilitation- It is the restoration of the previous land functions of the degraded land.
3. Land improvement- It is an increase in the environmental function or productive potential of land.

The following approaches are considered necessary for the wasteland reclamation-

i. Agroforestry
ii. Afforestation
iii. Reforestation
iv. Providing surface cover crops
v. Mulching
vi. Strip farming/Terracing/Contour ploughing
vii. Changing agricultural practices- crop rotation, crop pattern, mixed crop etc.
viii. Watershed management
ix. Soil conservation
x. Integrated management of natural resources
xi. Improving agricultural production
xii. Promotion of sustainable land management techniques to climate change adaptation
xiii. Adding value to agricultural, forestry, farmed products
xiv. Awareness raising, training and extension

Wasteland development schemes

Till the sixth five year plan, no specific programme of wasteland development was taken up. Government of India has set up the National Wastelands Development Board (NWDB) in 1985 under the Ministry of Environment & Forests, that the problem of wasteland development received a new thrust. With the setting up of NWDB, a number of new schemes were initiated to secure people's participation, besides continuation of ongoing afforestation schemes.

These are: Grants-in-aid to voluntary agencies:

- Decentralized People's nurseries
- Silvipasture farms
- Seed development
- Area oriented fuel wood and fodder projects
- Arial seeding programme
- Plantation of minor forest produce
- Margin money schemes
- Rural employment scheme

Major ongoing project

Integrated Wastelands Development Project (IWDP) Scheme

This scheme is under implementation since 1989-90, and has come to this Department along with the National Wastelands Development Board. The development of non-forest wastelands is taken up under this Scheme.

The major activities taken up under the scheme are:

- In situ soil and moisture conservation measures like terracing, bunding, trenching, vegetative barriers and drainage line treatment.
- Planting and sowing of multi-purpose trees, shrubs, grasses, legumes and pasture land development.
- Encouraging natural regeneration.
- Promotion of agro-forestry & horticulture.
- Wood substitution and fuel wood conservation measures.
- Awareness raising, training & extension.
- Encouraging people's participation through community organization and capacity building.
- Drainage Line treatment by vegetative and engineering structures
- Development of small water Harvesting Structures.
- Afforestation of degraded forest and non forest wasteland.
- Development and conservation of common Property Resources.

Some important organizations

CAZRI- Central Arid Zone Research Institute, Jodhpur, Rajasthan

CRIDA- Central Research Institute for Dryland Agriculture, Hyderabad, Telangana

FAO- Food and Agriculture Organization, Rome, Italy

GEF-Global Environment Facility, Washington DC

ICARDA- International Center for Agricultural Research in Dry Area- Aleppo, Syria (Beirut, Lebanon)

IWDP- Integrated Wasteland Development Programme

NWDB- National Wasteland Development Board, New Delhi

REDD- Reducing Emission from Deforestation and Forest Degradation

UNCCD-United Nations Convention to Combat Desertification, Bonn, Germany

UNDP- United Nations development Programme, New York

WSSD- World Summit on Sustainable Development, Johannesburg, South Africa

References

ENVIS Centre on Plants and Pollution, 2012. Waste Land Management through Plants, Ministry of Environment & Forests, Govt. of India.

Integrated Wastelands Development Programme, Department of Land Resources, Ministry of Rural Development, Govt. of India. https://dolr.gov.in/integrated-wasteland-development-programme

NRSA. 2005. Wasteland Atlas of India. Ministry of Rural Development and NRSA Publ., NRSA, Hyderabad.

Trivedi, T. P. 2010. Degraded and Wastelands of India Status and Spatial Distribution. Indian Council of Agricultural Research, Krishi Anusandhan Bhavan I, Pusa, New Delhi, pp. 1-167.

3

Area of Distribution of Wasteland and Problem Soils in Jharkhand

Deo Kumar

Introduction

Soil is a very precious natural resource. It is considered as the epicentre of all the life forms on earth. One cm of soil takes about 1000 years to form but if not used sustainably and managed properly, it may degrade and get converted into a wasteland in just few countable years. The heavy population load, rapid industrialization, unscientific agricultural practices have already put a lot of pressure on soil resulting in reduced soil fertility, degraded wastelands and problematic soils. In India, out of the 329 m ha of geographical area, nearly 50% is subjected to degradation by the various agents like water, wind and biotic interference. In this chapter we will focus on the wastelands and problematic soils, their distribution, cause of formation, management practices mainly in respect to the Indian state of Jharkhand.

Jharkhand is a state which was carved out of Bihar and was declared a new state on 15th of November, 2000. It has a total geographical area (TGA) of 79706 sq. km and extends from 28^0 58' N to 25^0 20' N latitude to 83^0 20'E to 87^0 57' E longitude on the map of India. It is a tribal dominated state, rich in mineral deposits and forest areas. It is a state with diversified land use patterns, major of which are urban and rural built Ups, mining, industrial areas, agricultural, forest lands, rivers, canals,reservoirs, wastelands etc.

Wasteland

Wasteland is a very broad term and since long history, a number of attempts have been taken by many authors to define it differently. According to Dudley Stamp (1954), "Wasteland is that land which has been previously used, but which has been abandoned, and for which no further use has been found". Wasteland Survey and Reclamation Committee, Ministry of food and Agriculture, Government of India (1961) has defined wastelands as those lands which are either not available for cultivation or left out without being cultivated for some reason or other. Bhumla and Khare (1987) gave a working definition of wastelands. According to them "Those lands are wastelands (a) which are ecologically unstable, (b) whose top soil has nearly been completely lost, and (c) which have developed toxicity in the root zones or growth of most plants, both annual crops and trees". A definition of wasteland givenby the National Remote Sensing Centre (NRSC) is "Wasteland is the degraded land which can be reclaimed with reasonable effort, and which is currently under-utilized andland which is deteriorating for lack of appropriate water and soil managementor on account of natural causes."

Causes of wasteland formation

Some of the major factors which are considered responsible for wasteland formation are:-

1. Deforestation
2. Overgrazing
3. Industrialization
4. Unscientific methods of crop cultivation and irrigation
5. Water logging
6. Salinity
7. Unplanned dumping of mining and urban wastes
8. Over exploitation of natural resources

Assessment of degraded and wastelands

Due to excess nutrient mining, adverse effects of climate change, excess accumulation of toxic elements in arable soils, their conversion into degraded and wastelands is a very serious concern for the land users. In mid sixties, some scientists who were working on wastelands and their management felt the need to classify the wastelands on the basis of their intrinsic characteristics like soil depth, soil texture, pH values, slope, erosion status and other inhibiting factors like salts, water-logging, rockiness, stoniness, etc. Since then, many attempts have been taken by different organizations to categorize it into suitable categories, among which NRSC emerged as the best. In India, the assessment of wastelands and mapping of their extent is done by NRSC (formerly known as National Remote Sensing Agency, NRSA) at the instance of Department of Land Resources (DoLR), Ministry of Rural Development (MRD), Government on India (GOI) in collaboration with various State and Central Departments.

GOI publishes "Wastelands Atlas of India" after each assessment process is completed and results are drawn out of it. Till date, government has released 4 atlases in the years 2000, 2005, 2010 and 2011 to have a regular view and control on the extent of wastelands.

Mapping techniques

Mapping is the process of recording the findings of assessment process in a pictorial form under a specified scale. Mapping is very necessary as the maps serve many purposes like studying the wastelands, planning their management or utilization. In India, primarily the following two systems of mapping are followed:

1. Conventional System
2. Aerial photography and Remote sensing

1. Conventional system

This system of mapping is a very time taking and labour intensive system as it involves ground transacts using base maps. Transact walks are generally done with the help of topographical maps of scale 1:50000, prepared by Survey of India. The surveyor walks through the wastelands and collects data by observing visual characteristics and collecting soil and land use data and use it for mapping.

2. Aerial photography and remote sensing

These are the modern techniques used for mapping a large area very simply in a very short time. This is done by capturing images of the earth's surface using sensors which are mounted on different types of platforms namely weather balloons, UAVs, satellites, etc. After capturing images, False Colour Composites (FCCs) are prepared out of them and interpretations are done. It is a more efficient way of mapping wastelands as the orchards, croplands, forest areas and other similar looking features can be segregated using their specific spectral reflectance. This method of mapping is quicker and cost effective as compared to conventional method.

NRSC uses remote sensing technologies to assess and evaluate the extent of wastelands in India. Multi temporal satellite images are used for this purpose. For the preparation of recent wasteland atlas i.e Wastelands Atlas of India, 2011, 3 season (kharif, rabi and zaid seasons of year 2008-09) multi temporal satellite images of Resourcesat-1 LISS III at a scale of 1:50000 were used and wasteland maps at country as well as state and union territory level was released.To evaluate the change in total area of wastelands, NRSC observed the spatial changes in wastelands between 2005-06 and 2008-09. In The wasteland Atlas of 2010, a total wasteland area of 47.22 M ha was reported under 23 categories of wastelands for the year 2005-06 and this area reduced to 46.7 M ha in the 2011 atlas. In The Wasteland Atlas of 2011, main focus was laid on the change analysis of wasteland categories during 2005-06 and 2008-09 for the whole country on 1:50000 scale. In this analysis, a decrease of 3.2 M ha of previous wastelands was observed but simultaneously, an increase of 2.7 M ha in form of new as well as extension of old wastelands was also observed thus, making a total decrease of 0.5 M ha of wasteland from the 2010 data.

Classification of wastelands

Department of Land Resources (DoLR), Ministry of Rural Development (MRD), Government on India (GOI) follows the classification system which was suggested by NRSC in "Wastelands Atlas of India 2010". Under this classification the total wastelands have been classified under 23 categories, which have again been grouped into 8 major classes, shown in Table 3.1.

Table 3.1: Classification of Wastelands in India

	Sl. No.	Wastelands Category		Sl. No.	Wastelands Category
A		Gullied/Ravinous land	F		Scrub Forest (Under utilized notified forest land)
	1	Medium ravine		11	Scrub dominated
	2	Deep/very deep ravine		12	Agricultural land inside notified forest land
B		Scrubland (Land with or without scrub)		13	Degraded pastures/grazing land
	3	Land with dense scrub		14	Degraded land under plantation crops
	4	Land with open scrub	G		Sands(coastal/desert/riverine)
C		Waterlogged and Marshy land		15	Sands- Coastal sand
	5	Permanent		16	Sands- Desert sand
	6	Seasonal		17	Semi-stabilized to stabilized (>40m) dune
D		Land affected by salinity/ alkalinity		18	Semi-stabilized to stabilized moderately high (15-40m) dune
	7	Moderate		19	Sands-Riverine
	8	Strong	H		Mining/Industrial wastelands
E		Shifting Cultivation		20	Mining wasteland
	9	Current Jhoom		21	Industrial wasteland
	10	Abandoned Jhoom		22	Barren rocky area
				23	Snow cover and/or glacial area

1. Gullied and ravinous lands

These are the hilly/mountainous terrians having steep slopes, which are prone to water erosion. Through the action of water 4 types of erosions occur namely, splash, sheet, gully and rill erosion. Gully and rill erosion are the advance form of water erosion in which initially a series of small channels are formed in soil by the running water which gradually widens and form an intricate network of large and deep gullies known as ravines. Formation of gullies in soil is governed by many soil and climatic factors like soil cover, soil texture, rainfall intensity and duration, slope length etc.In the satellite images, ravines are delineated into two categories viz. medium ravines and deep ravines based on their depth.

1.1 Medium ravines

The ravines of relatively shallow depth ranging from 2.5m to 5m are kept under this category. They are generally seen near to the head region of the stream, mainly close to agricultural land from where the erosion starts due to movement of the flowing stream. The ravines are generally shallow in this area because the stream current is generally slow at its origin and gains velocity as it moves ahead.

1.2 Deep ravines

This category consists of the ravines which generally are deeper than 5m. These ravines are found along the higher order stream areas that are close to the main river. The flowing stream gains velocity as well as erosive power along with its further movement thus results into deeper ravines.

2. Scrubland

These are the lands found in topographically high locations, other than hilly/ mountainous terrains, which are prone to deterioration by erosion. The soil depth in these areas is generally shallow due to some or other constrains related to the soil forming factors. As the density of vegetation cover over an area can be easily delineated on a satellite image, it has been taken as a criterion of classification of its sub classes. On this basis, these areas have been delineated into two sub-classes viz. land with dense scrub and land with open scrub. The land with dense scrub is generally covered with dense vegetation cover and on a satellite image, it is somewhat difficult to differentiate such land from croplands due to similar spectral reflectance shown by both types of land. The characteristics of the second class i.e the land with open scrub is similar to that of the land with dense scrub except that the vegetation cover in this class is sparse.

3. Waterlogged/Marshy land

The land which is generally low lying and stay inundated with water for most of the year is known as waterlogged soil. The standing water over the land changes its physical, chemical as well as biological properties which in turn alters the ability of the soil to supply nutrients in balanced amount and ratio, and makes the soil unfit to grow crops. On the basis of fact whether the soil remains covered with standing water throughout the year or dries for at least some part of the year, this land is delineated into 2 sub-classes viz. permanently waterlogged and seasonally waterlogged areas. The low lying area having impervious substratum where waterlogged condition prevails throughout the year, is known as permanently waterlogged soil whereas the land lying in plain area but having poor drainage conditions, and generally remain waterlogged only during the monsoon season is known as seasonally waterlogged area.

4. Land affected by salinity/alkalinity

The accumulation of dissolved salts in the surface layer of a low rainfall area, where the evapo-transpiration is more than that of the precipitation and the rainfall is insufficient to leach the dissolved salts from surface to sub-surface layers is known as soil salinity. Generally encrustation of salts on the surface layer can also be seen.This is due the upward capillary movement of dissolved salts along with rising water from sub-surface layers as a result of high evapo-transpiration. This is a harmful condition for the plants growing in that area because the dissolved salts increase the osmotic potential of the soil. Due to increased osmotic potential in the rhizosphere, even though the water is physically present, due to development of a negative pressure gradient the roots are unable to take soil water and the plants suffer from a water stress condition. This condition is known as physiological drought. Alkaline soils are those in which the soluble salts are dominated with sodium ions resulting into high ESP value. It deteriorates the physical condition of soil by causing dispersion and destroying the soil structure. After destruction of the structure, the fine particles block the micro and macro pores present in soil preventing the percolation of water and creating a waterlogged condition. According to the degree of salinity and or alkalinity, and some other properties, the salt affected land has been delineated into two sub-classes viz. moderately saline/alkali and strongly saline/alkali areas.

4.1 Moderately saline/Alkali land

The land whose Electrical Conductivity (EC) ranges between 8 to 30 dS/m, pH ranges between 9.0 to 9.8 and the Exchangeable Sodium Percentage (ESP) value ranges between 15-40, has been kept under this category.

4.2 Strongly Saline/Alkali land

The land whose EC, pH and ESP value exceeds the range of moderately saline/alkali land has been categorized under this category. The EC value of strongly saline/alkali soil is greater than 9 dS/m, pH value is more than 9.8 whereas the ESP value is more than 40.

5. Land affected by shifting cultivation

Shifting cultivation, which is also known as Jhoom cultivation or Slash and Burn Cultivation is a traditional practice of cultivation of crops on hill slopes and forested area, where a part of land is selected and the prevailing vegetation is cleared by cutting it down and incorporating it in soil in form of ash after burning it. Then, cultivation of crops is done for few years on that cleared part of the land until its fertility status goes down and it becomes unable to give high yield. When the farmers don't get profitable yield from that land, they abandon it and shift to a fresh piece of forest/hilly land, clear it and start cultivation of crops there. These wastelands have been classified under 2 sub-classes namely Current Shifting Cultivation Areas and Abandoned Shifting Cultivation Areas.

5.1 Current shifting cultivation areas

These are the areas where presently shifting cultivation is going on. These areas can be easily recognized and separated in a satellite image by its pre-burnt and post-burnt conditions.

5.2 Abandoned shifting cultivation areas

These areas are the ones where shifting cultivation was once being practiced but now left abandoned for more than 1 year but less than 5 years as they became infertile and are unable to give expected yield. They can be separated out from the rest of forest areas by looking at the regenerated secondary vegetations such as tall grasses, shrubs, bamboo, etc. If they are left abandoned for more than 5 years, it becomes very difficult to separate them from the rest of forest due to very dense vegetative growth.

6. Scrub forest land

The forest lands which are underutilized having sparse vegetation dominated by shrubs, often also including grasses, herbs, and geophytes are called scrub forest. Sometime the scrub forest also include agricultural lands in between them, so they have also been classified under this category. Scrub Forest land has been delineated into 2 sub-classes viz., degraded forest land dominated by scrub and agriculture land inside notified forest area.

6.1 Degraded forest land dominated by scrub

In the Forest act, it has been notified that if a forest land has less than 20 percent of its area covered by vegetative cover, it is classified as a degraded forest. These degraded forests are mainly dominated by scrubs and found in the fringe areas of a notified forest.

6.2 Agriculture land inside notified forest land

Many a times it has also been observed that some part of the notified forest area is brought under cultivation of agricultural crops by some of the local inhabitants or tribal people living near to the forest. This category of land has also been considered as a category of wasteland by NRSC as that part of land has the capacity to support much more vegetation but is currently underutilized.

7. Degraded pastures/Grazing land

When the pastures are over utilized or cattle are allowed to graze on a same piece of grazing land repeatedly, without providing it enough time gap to recover from its former grazed state, it is called overgrazing. It is an environmental hazard where, cattle excessively graze on the pasture causing the removal of plant cover from soil surface which results into exposure of the bare land to the agents of erosion and contribute to its degradation. Its development mainly occurs due to poor soil management techniques, carelessness and improper soil conservation techniques.

8. Degraded land under plantation crop

These are the lands which were once degraded but now by the interventions of government and/or local people living over there, reclamation techniques have been adopted and applied in the form of plantation crops. These lands are generally located outside the notified forest areas. Here, such plants are grown, which are hardy in nature, can grow in poor soils, establish easily, grows quickly and require little care.

9. Sand (coastal/desert/riverine)

Sand alone cannot support plant growth as it lacks the property of cohesion and adhesion and thus is unable to hold and supply the nutrients and water to plants. Thus, the sandy lands are considered as wastelands. These lands are generally delineated in the sea shores, deserts and river beds. The coastal sands are found along the seashores and are accumulated by the action of tides. The desert sands are found in the aridic environment where precipitation is very less. Its transportation takes place through the force of blowing winds and known as aeolian transportation. When the force of blowing wind is more, its carrying capacity is also more thus capable of carrying sand with it, but as its force starts declining, it cannot sustain the weight of carried sands and the sands are deposited forming dunes of varying size. The height of the dunes keeps on changing regularly due to continuous removal and deposition of sand from a particular place. Based on the vertical approximate height of sand dunes, they are classified into 2 categories viz.

i) semi-stabilized to stabilized dunes with less than 40m height

ii) semi-stabilized to stabilized moderately high dunes with heights ranging between 15 m and 40 m.

Lands dominated by the riverine sands are found besides the river as flood plains. These are the sands deposited in form of sheets by rivers during its course of movement.In some parts of the Indo-Gangetic plains inland sand is also found which has accumulated in the abandoned river courses which once used to flow from that area or due to the reworking of sand deposits by wind actions.

10. Mining/industrial wastelands

In the mineral rich parts of the earthlands, these sections of the wastelands are found as debris which has accumulated due to the extraction process around the mineral areas. Though these lands are initially fertile and productive, they become wastelands either due to excavation of mining pits or due to getting covered with the mining and industrial wastes. Sometime they may be productive but due accumulation of heavy and harmful metals, agriculture is not practiced in such areas.

11. Barren rocky /Stony waste

Some part of the earth surface is either rocky or contains a very thin soil layer whereas some part, due to mismanagement, lose the important soil separates leaving only stones on surface. Such wastelands are known as rocky or stony wastelands.Cultivation of crops in these types of lands is very difficult because either they are devoid of soil or the soil depth is not so deep that it can sustain plant life.

12. Snow covered / Glacial area

Some areas, especially those lying in the Himalayan regions, remain covered with snow for a significant period of time. Being at very low temperature and covered with snow, these lands cannot be utilised. Thus, they have been delineated as wasteland. Though their extent is limited only to few states but cover a significant area on the map.

Distribution of wastelands in India

A nationwide mapping of wastelands in India was carried out by NRSC at the behest of Department of Land Resources (DoLR), Ministry of Rural Development (MRD), Government of India. Under this project, NRSC mapped the whole area of country at district level leaving an area of 12.08 M ha in Jammu & Kashmir and released wasteland map of India at scale 1:50000 showing the extent of different categories of wastelands in different states of Indias. The wasteland map showed that out of 316.64 M ha of total mapped area, 46.702 M ha (14.75 % of TGA) are wastelands, shown in fig 3.1.

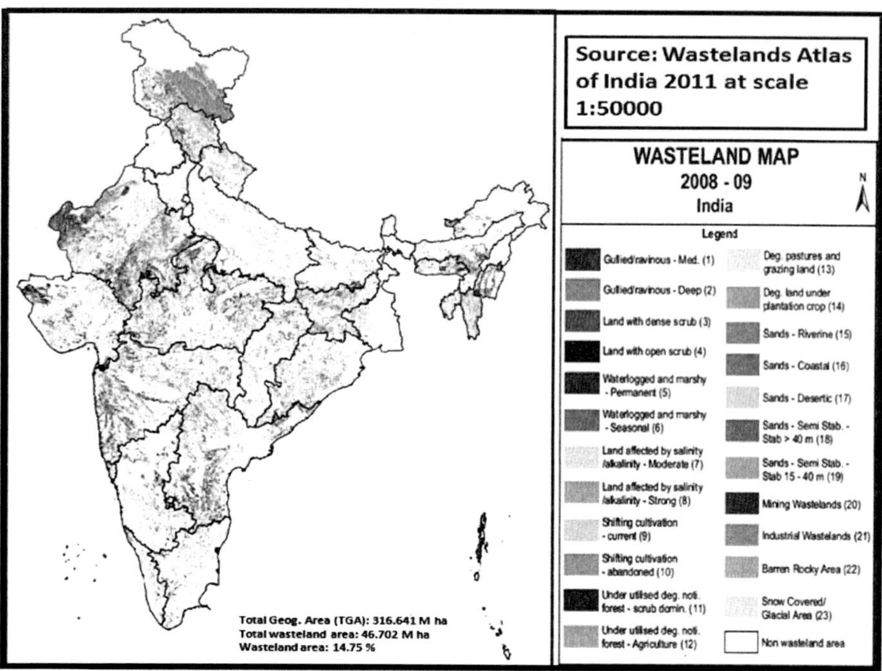

Fig. 3.1: Wastelands Map of India (*See colour plate on page 271*)

Table 3.2: State wise distribution of Wastelands

Sl.No.	State Name	No. of Districts	Total Geographical Area (M ha)	Total Wastelands Area (M ha)
1	Andhra Pradesh	23	27.507	3.723
2	Arunachal Pradesh	16	8.374	1.490
3	Assam	23	7.844	0.845
4	Bihar	37	9.417	0.960
5	Chhattisgarh	16	13.519	1.148
6	Delhi	1	0.148	0.009
7	Goa	2	0.370	0.049
8	Gujarat	25	19.602	2.011
9	Haryana	21	4.421	0.215
10	Himachal Pradesh	12	5.567	2.235
11	Jammu & Kashmir	14	10.138	7.544
12	Jharkhand	24	7.971	1.102
13	Karnataka	27	19.179	1.303
14	Kerela	14	3.886	0.245
15	Madhya Pradesh	48	30.825	4.011
16	Maharashtra	35	30.769	3.783
17	Manipur	9	2.233	0.565
18	Meghalaya	7	2.243	0.413
19	Mizoram	8	2.108	0.496
20	Nagaland	7	1.658	0.527
21	Orissa	30	15.571	1.643
22	Punjab	20	5.036	0.937
23	Rajasthan	32	34.224	8.493
24	Sikkim	4	0.710	0.327
25	Tamil Nadu	30	13.006	0.872
26	Tripura	4	1.049	0.096
27	Uttarakhand	13	5.348	1.286
28	Uttar Pradesh	70	24.093	0.988
29	West Bengal	19	8.875	0.193
30	Union Territory	8	0.949	0.032
	Total	599	316.64	46.702

Source: Wastelands Atlas of India, 2011.

Rajasthan due to its arid climatic conditions and almost negligible vegetative cover, constitutes the maximum area under wastelands among all the states and union territories followed by Madhya Pradesh and Maharashtra. Due to undulating and mountainous topography and the action of agents of erosion like wind and water, the states of Andhra Pradesh, Gujarat, Himachal Pradesh, Jammu & Kashmir, Jharkhand, Madhya Pradesh, Maharashtra, Orissa, Rajasthan, Tamilnadu and Uttar Pradesh are severely affected by the Gullied and ravinous land category of wasteland,. Scrublands are the major problem of almost all the states. The major extent of waterlogged and marshy lands lies in the states of Andhra Pradesh, Assam, Bihar and Orissa with a little extent in Tamil Nadu, Rajasthan, Punjab, Maharashtra, Jammu & Kashmir and Goa. Salinity and alkalinity develop in the arid regions, where the evapo-transpiration is more than precipitation and the coastal areas which remain submerged in brackish water for most of the time of a year. Most of the saline and alkaline lands are mainly found in the states of Uttar Pradesh, Gujarat, Karnataka, Andhra Pradesh and Tamil Nadu having some spread in the states of Orissa, Punjab, Maharashtra, Jammu & Kashmir and Haryana. The practice of shifting cultivation is still delineated in some of states like Orissa and the north eastern states of Arunachal Pradesh, Assam, Manipur, Meghalaya, Mizoram, Nagaland and Tripura. The forest lands of more or less, all the Indian states are under-utilized either in form of scrub dominated or agricultural land within the notified forest lands.

Minerals are very important for the economic prosperity of a state but it also has a major drawback that while extraction of the mineral from deep earth, a vast wasteland is also created. This category of wasteland is mainly found in the states of Rajasthan, Tamilnadu, Madhya Pradesh, Jharkhand, Karnataka, Haryana, Maharashtra, Uttar Pradesh, West Bengal and in little extent to some other states. The total area of industrial wastelands throughout the country is 5800 hectares, which is least in comparison to all other categories of wastelands. The states having the major extent of industrial wastelands are Uttar Pradesh, Rajasthan, Orissa, Andhra Pradesh, Bihar, Chhattisgarh, Jammu & Kashmir, Jharkhand, Madhya Pradesh and Tamilnadu. The barren rocky or stony wastelands is also a very important category of wastelands as it has got vast extents almost throughout the country having its major areas of spread in the states of Jammu & Kashmir, Rajasthan, Andhra Pradesh, Himachal Pradesh, Uttarakhand, Maharashtra and Karnataka. Some parts of the states like Jammu & Kashmir, Himachal Pradesh, Uttarakhand, Sikkim and Arunachal Pradesh, where the Himalayan ranges are situated, remain covered with snow and glaciers, which form an another category of wastelands. The state-wise extent of wastelands has been depicted in Table 3.

Table 3.3: Category wise total area under wastelands in India

Sl.No.	Category of Wasteland	Wasteland Area (M ha)
1	Gullied and/or ravenous land Medium	0.615
2	Gullied and/or ravenous land-Deep/very deep ravine	0.127
3	Land with dense scrub	8.698
4	Land with open scrub	9.303
5	Waterlogged and Marshy land- Permanent	0.176
6	Waterlogged and Marshy land- Seasonal	0.695
7	Land affected by salinity/alkalinity-Moderate	0.541
8	Land affected by salinity/alkalinity-Strong	0.139
9	Shifting cultivation area-Current Jhum	0.481
10	Shifting cultivation area-Abandoned Jhum	0.421
11	Under utilised/degraded forest-Scrub dominated	8.370
12	Agricultural land inside notified forest land	1.568
13	Degraded pastures/grazing land	0.683
14	Degraded land under plantation crops	0.028
15	Sands-Riverine	0.211
16	Sands-Coastal sand	0.065
17	Sands-Desert sand	0.393
18	Sands-Semi-stabilized to stabilized (>40m) dune	0.928
19	Sands-Semi-stabilized to stabilized moderately high (15-40m) dune	1.427
20	Mining Wastelands	0.059
21	Industrial wastelands	0.006
22	Barren rocky area	5.948
23	Snow cover and/or glacial area	5.818
	Total	**46.702**

Source: Wastelands Atlas of India, 2011.

From the above given data of the category wise distribution of wastelands throughout India, it can be concluded that a dominant part of wastelands lie in Table 3.3 under scrublands. The two categories of scrublands viz. land with open scrub and land with dense scrub cover area of 9.303 M ha and 8.698 M ha respectively across the country. They collectively constitute an area of 18.001 M ha, which is approximately 38.5 percent of the total wastelands area of India. Besides these two categories, barren rocky area and snow covered or glacial covered lands have also major extent in the country. Collectively they cover an area of nearly 25 percent of the total Indian wastelands. Among all the categories of wastelands in India, the extent of mining and industrial wastelands are least.

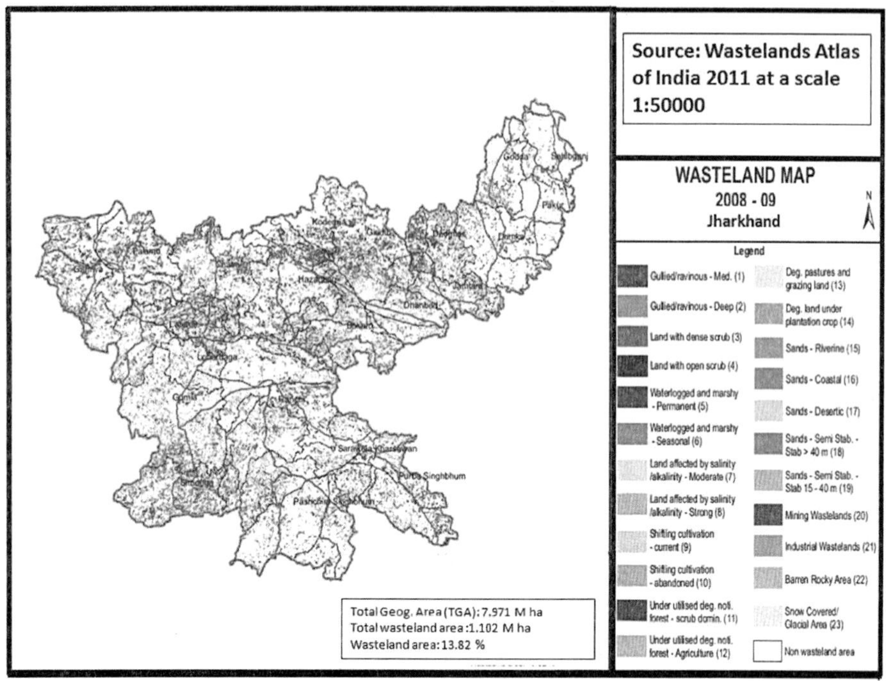

Fig. 3.2: Distribution of Wastelands in Jharkhand *(See colour plates on page 272)*

Wasteland map of Jharkhand

In the wastelands map of Jharkhand, released in "Wastelands Atlas of India 2011" shown in fig. 3.2, out of the total geographical area of 7.971 M ha, nearly 1.1 M ha is delineated as wastelands, which is approximately 13.82 percent of its total geographical area. Out of 23 classified categories, only 11 have been reported in Jharkhand among which, 8 are major and the rest 3 are very less in extent. Under-utilized forest lands dominated by scrubs have the maximum extent in Jharkhand followed by wastelands with dense and open scrub. These 3 most dominated categories prevail in all the 24 districts, shown in table 3.4 & 3.5. Medium gullied and ravinous wastelands are majorly distributed in the districts Chatra, Hazaribagh, Gumla, Latehar and Kodarma. Out of the total area of 29,335 Ha under mining wastelands, major fall in the districts of West Singhbhum, Ranchi, Ramgarh, Hazaribagh and Deoghar. Small extents are also seen in the districts Simdega, Pakur, East Singhbhum, Dhanbad, Chatra, Lohardaga, Godda, Kodarma, Lohardaga and Palamau. The barren rocky wastelands are also found in all the districts of Jharkhand, majorly in Gumla, Garhwa, Palamau, Ranchi, Simdega and West Singhbhum. Very little area of waterlogged and marshy wastelands are found in the districts of Dhanbad, Pakur, Sahibganj and Saraikela-Kharswan. Industrial wastelands have been delineated in a few districts like Bokaro, East Singhbhum, Ramgarh, Sahibganj and West Singhbhum, due to the presence and development of industries in these areas.

Table 3.4: District-wise total wastelands in Jharkhand

Sl.No	District	Total Wastelands Area
1	Bokaro	375.97
2	Chatra	719.04
3	Deoghar	610.18
4	Dhanbad	186.10
5	Dumka	277.63
6	East Singhbhum	430.64
7	Garhwa	676.28
8	Giridih	1040.41
9	Godda	242.52
10	Gumla	655.88
11	Hazaribagh	732.96
12	Jamtara	196.33
13	Kodarma	189.47
14	Khunti	274.43
15	Latehar	778.33
16	Lohardaga	86.62
17	Pakur	57.23
18	Palamau	669.98
19	Ramgarh	258.80
20	Ranchi	540.62
21	Sahibganj	104.75
22	SaraikelaKharswan	297.27
23	Simdega	940.54
24	West Singhbhum	675.38
	Total	11017.38

Table 3.5: District-wise distribution(in hectares) of different categories of Wastelands in Jharkhand.

Sl. No	WL Category District	1	2	3	4	5	6	7	8	9	10	11	12	13	14	15	16	17	18	19	20	21	22	23	Total Area (Ha)
1	Bokaro	214		6016	11502							13479	5766								274	7	340		37597
2	Chatra	5821		1954	14038							43260	6393								34		404		71904
3	Deoghar	884		10001	33994							10547	5365								121		105		61018
4	Dhanbad	200		6225	5147	10						6031	667								34		295		18610
5	Dumka	264		5312	12844							8510	630			2							201		27763
6	E. Singhbhum			14839	9236							17507	491								65	26	900		43064
7	Garhwa	753		16574	7486							38746	1727										2342		67628
8	Giridih	787		3098	36242							472.96	15964										654		104041
9	Godda	22		14203	5128							3777	532								3.34		257		24252
10	Gumla	2198		25454	7360							16987	4529										9061		65588
11	Hazaribagh	4042		8359	14628							432.74	1641								360		992		73296
12	Jamtara	182		2187	11302							5041	66										855		19633
13	Kodarma	1323		3721	6435							6755	587								2		124		18947
14	Khunti	34		11396	5760							7004	2797										452		27443
15	Latehar	1064		23709	13615							39054	359										32		77833
16	Lohardaga	614		800	3563							2895	368								4		419		8662
17	Pakur	55		4228	616	4						396	346								32		48		5723
18	Palamau	519		8021	9155							43521	2894								7		2881		66998
19	Ramgarh	151		8631	7036							8162	815								451	9	624		25880
20	Ranchi	16		162.29	80.58							22389	3174								431		3765		54062
21	Sahibganj			3813	2212	26	29					3472	641									20	261		10475

Sl. No	WL Category District	1	2	3	4	5	6	7	8	9	10	11	12	13	14	15	16	17	18	19	20	21	22	23	Total Area (Ha)
22	Saraikela-Kharswan			8743	11042		2					9112	449										379		29727
23	Simdega	4		46009	4168							38311	2941								76		2545		94054
24	W. Singhbhum			31650	11007							20016	2953								511	1	1399		675.38
	Total	19147		281169	251575	41	31					455544	62094			2					2737	62	29335		1101738

Wastelands categories

1. Gullied and/or ravenous land Medium
2. Gullied and/or ravenous land-Deep/very deep ravine
3. Land with dense scrub
4. Land with open scrub
5. Waterlogged and Marshy land- Permanent
6. Waterlogged and Marshy land- Seasonal
7. Land affected by salinity/alkalinity-Moderate
8. Land affected by salinity/alkalinity-Strong
9. Shifting cultivation area-Current Jhum
10. Shifting cultivation area-Abandoned Jhum
20. Mining Wastelands
22. Barren rocky area

(*Source*: Wastelands Atlas of India, 2011.)

11. Under utilised/degraded forest-Scrub dominated
12. Agricultural land inside notified forest land
13. Degraded pastures/grazing land
14. Degraded land under plantation crops
15. Sands-Riverine
16. Sands-Coastal sand
17. Sands-Desert sand
18. Sands-Semi-stabilized to stabilized (>40m) dune
19. Sands-Semi-stabilized to stabilized moderately high (15- 0m)dune

21. Industrial wastelands
23. Snow cover and/or glacial area

Wasteland development

In mid 80s, many of the Indian states were suffering from ecological imbalance and crisis of fodder and fuel wood. To overcome these situations, Government of India formed National Wasteland Development Board in 1985 so that the wastelands could be brought under utilization and help in mitigating the prevailing problems. It mainly focused on peoples' participation for developing the wastelands so that the land degradation could be checked and the wastelands could be utilized to some extent increasing the biomass of the area. The main steps involved in wasteland development are:-

1. **Afforestation-** Though cultivation is not an easy task in wastelands, but if proper species of trees are selected for particular types of wastelands, they can easily grow and increase their biomass, restoring the ecological balance. Tree species like *Acacia auriculiformis, Zizyphus maurtiana, Eucalyptus camaldulensis, Bambusa spp., Tamarindus indica,* etc. can easily grow under different harsh conditions. Their root holds the soil and prevent its erosion due to runoff whereas canopy prevents splash erosion caused due to the beating action of raindrops.

2. **Increasing the surface vegetative cover**-Different methods like mixed cropping, rotational cropping, cover cropping, etc. are practiced in order to conserve soil, providing it stability and prevent from erosion. The cover crops and strip planting techniques, conserves the soil and prevent erosion by a number of activities. Firstly, it prevents the direct fall of raindrops on the soil, which is reason of splash erosion. Secondly, it does not allows the free surface runoff, preventing increase in its carrying capacity. Thirdly, it not only holds and provides strength to the soil by help of its fibrous and tap roots, but also adds to the organic matter, which improves soil's physical as well as chemical health.

3. **Mulching-**A protective cover of different plant parts like leaves, straws, hay or other dead and decaying organic matter should be applied over the earth's surface in order to reduce soil erosion, reduce evaporation and retain moisture in the soil. The organic matter added to the soil also helps to build up the soil health. Thus it is always recommended to leave some crop residues in the field at the time of harvesting.

4. **Contour Farming, Terracing and Trenching-**In steep slope areas, where the speed of runoff is sufficiently high to cause soil erosion, agronomical and engineering methods like strip cropping, contour farming, and contour trenching are practiced. The main motive here is to hinder the path of the water flow so that its carrying capacity stays in limit and does not cause soil erosion. In slightly sloped areas, crops are

grown in strips along a contour line whereas in steep slope areas, steps like platforms or plots are prepared and crop cultivation is carried out. In order to collect the eroded soil and store some amount of running water for later use, trenches are also dug at fixed distances.

5. **Watershed Development-**Watershed is an area bounded hydrological system, from ridge top to ridge bottom that collects all the precipitation of that area and releases water through different channels to a main channel and the water exits the system by a single (lowest) point. Though a watershed development requires a huge man power and planning, once developed, can sustain whole area and can support in the wasteland development by increasing biomass production of the area, increasing the forest cover and agricultural produce, making the area diversified in terms of different species dwelling there.
6. **Development of recreational areas-**Now a days, wastelands are also being converted into recreational areas like parks. It's best example can be seen in Jugsalai Eco park, Jamshedpur, Jharkhand where a fly ash hill formed by the continuous dumping of fly ash, a waste product of Tata steel industries, has been transformed into an Eco park.
7. **Utilization of mining pits as water storage-**After mining, deep pits are left over, which can be utilized for storing the rainwater during monsoon, which may be used later to irrigate the plantation sites and croplands, during summer season. This may reduce the heavy load on underground water during summer and lead to water conservation.
8. **Topsoil Replacement**-The productivity of rocky barren areas or other wastelands may be increased by replacing or covering its topsoil by some fertile soil. The fertile topsoil enhances the nutrient status as well as physical characteristics of the soil which in turn increases the water retention capacity, buffering capacity, proper root growth etc. which directly affects the crop growth.

References

Anonymous, 2000. Wasteland Atlas of India, (2010). Ministry of Rural Development, Govt. of India and National Remote Sensing Agency (NSRA), Dept. of Space, Govt. of India, Hyderabad.

Anonymous, 2010. Wasteland Atlas of India, (2011). Ministry of Rural Development, Govt. of India and National Remote Sensing Agency (NSRA), Dept. of Space, Govt. of India, Hyderabad.

Balasubramanian, A. 2015. The Wastelands In India. 10.13140/RG.2.2.35824.87045.

Maji, A. K., Reddy, G. O., and Sarkar, D., 2010. Degraded and wastelands of India: status and spatial distribution. Indian Council of Agricultural Research, New Delhi, p.167.

Singh, R. S. and Tripathi, N. 2005. Watershed Management in Mined-out Wasteland Areas. *Jharkhand Journal of Development and Management Studies XISS*, Ranchi, 3(2): 1411-1429.

4

Soil Forming Factors and Processes

Abhay Kumar, Bijay Kumar Singh and P. R. Oraon

Introduction

Pedology:- Pedology word drive from Greek word pedon, means soil or earth and logos means to study. Pedology is concerned with origin of the soil, its classification, distribution and character (Physical, Chemical and Biological).

Definition:- Soils are natural bodies made from both mineral and organic materials and capable of supporting plants out-of-doors.

Composition of soil on volume basis (Soil components)

i. Mineral matter :- 45%

ii. Organic matter :- 5%

iii. Soil water :- 25%

iv Soil air :- 25%

Soil formation is a long term process. It takes several million years to form a thin layer of soil. The soil is made up of broken down rock material of varying degree of fineness and changed in varying degrees from the parent rocks by the action of different agencies such that the growth of vegetation is made possible. As soil is a complex mixture of various components, its formation is also more complex. The Formation of a particular type of soil depends upon the physico-chemical properties of the parent rock, intensity and duration of weathering, climatic and other parameters.

Origin of soils

The force of wind, water or glaciers might have transported the soil to some other place. In addition to the soil parent material, origin of soil is also dependent upon other prevailing processes affecting soil formation. Climatic conditions are important factors affecting both the form and rate of physical and chemical weathering of the parent material. The formation of soils can be seen as a combination of the products of weathering, of structural development of the soil, of differentiation of that structure into horizons or layers, and lastly of its movement or translocation.

Soil forming

The evolution of true soil from regolith (weathering of rock) takes place by the combined action of soil forming factors and processes. The soil formation is the process of two consecutive stages.

- The first step is accomplished by weathering (disintegration & decomposition) of rock (rock into Regolith).
- The second step is associated with the action of Soil Forming Factors (Regolith into soil).

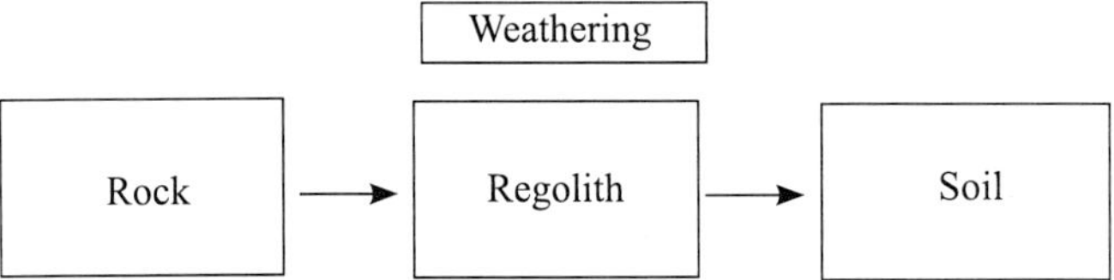

Soils are the products of weathering from some parent rocks. All soils initially come from some pre-existing rocks. They are called as 'parent materials'. The Parent Material may be directly below the soil or at great distances from it. It is necessary to understand the factors and processes that are responsible for the formation of soils. The complexity of soil patterns, texture, composition, and color in different areas highly depends on the physical and chemical compositions of the parent materials. Most soil parent materials were rocks at some time in their history. The minerals in rocks contribute to soil fertility and other soil properties long after the original rock is gone. Geologists classify rocks into igneous, sedimentary and metamorphic rocks, according to their origins. Weathering breaks up rock and minerals, changes their chemical composition and characteristics, and also synthesizes a soil from the upper part of the regolith. This unconsolidated material is acted on by the five soil forming factors and soil is formed.

Soil forms continuously, but slowly, from the gradual breakdown of rocks through weathering. The surface rocks are exposed to the process of weathering. In this process, the rocks are converted into fine grains and provide a base for the soil formation. Two types of minerals are found in natural systems: primary and secondary. Whether the mineral is primary or secondary depends on the mode of formation and not on the mineral composition. Minerals that crystallize from cooling magma are called primary and minerals that crystallize during the weathering of primary minerals are called secondary minerals.

Weathering can be a physical, chemical or biological process:

i. Physical weathering-breakdown of rocks from the result of a mechanical action. Temperature changes, abrasion (when rocks collide with each other) or frost can all cause rocks to break down.

ii. Chemical weathering-breakdown of rocks through a change in their chemical makeup. This can happen when the minerals within rocks react with water, air or other chemicals.

iii. Biological weathering-the breakdown of rocks by living things. Burrowing animals help water and air get into rock, and plant roots can grow into cracks in the rock, making it split.

Factors

Dokuchaiev (1889) established that the soils develop as a result of the action of soil forming factors

$S = f(P, Cl, O)$

Further, Jenny (1941) formulated the following equation

$S = f(Cl, O, R, P, T, \ldots)$

Where,

Cl = Climate

O = Organisms and vegetation (biosphere)

R = Relief or topography

P = Parent material

T = Time

...= additional unspecified factors

The five soil forming factors, acting simultaneously at any point on the surface of the earth, to produce soil.

Factors affecting soil formation

Pedogenesis can be defined as the process of soil development. Soil development is a function of climate, organic matter, relief, parent material and time. The accumulation of material through the action of water, wind and gravity also contributes to soil formation. These processes can be very slow, taking many tens of thousands of years. Five main interacting factors affect the formation of soil, shown in fig 4.1.

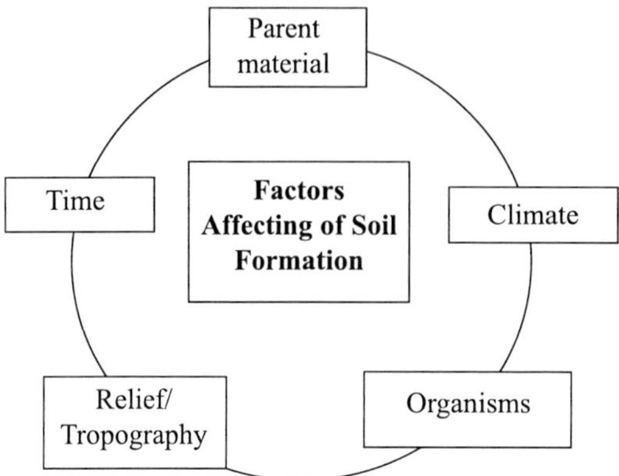

Fig. 4.1: Five main interacting factors affect the formation of soil

1. Parent Material- The rocks from which soils are formed are called parent materials. The parent material determines the colouration, mineral composition and texture of the soil. The mineral from which the soil is formed is termed as the parent material. Rocks are the source of all soil minerals. The parent material is chemically or physically weathered and transported which then deposits to form layers of soils.

Occurrence of soil forming rocks. The composition of the upper 5 km of the Earth crust is as follows:

Sedimentary rocks	- Shales 52 %
	- Sandstones 15 %
	- Limestones and dolomite 7 %
Igneous rocks	- Granite 15 %
	- Basalt 3%
Other rocks	- 8%

So it is found that five kinds of rocks mentioned above occupy more than 9O% of the total continental area.

In India, parent material is generally categorized into:

i. Ancient crystalline and metamorphic rocks- They are the oldest rocks (formed due to solidification of molten magma about 4 billion years ago). They are basically granites, gniesses and schists. These rocks are rich in ferromagnetic materials and give rise to red soils on weathering. The red colour of these soils is due to the presence of iron oxide.

ii. Cuddapah and Vindhyan rocks-They are ancient sedimentary rocks (4000 m thick).On weathering they give calcareous (containing calcium carbonate; chalky) and argillaceous (containing clay) soils.

iii. Gondwana rocks-These rocks are also sedimentary in nature and they are much younger. On weathering they give rise to comparatively less mature soils.

iv. Deccan basalts- Volcanic outburst over a vast area of the Peninsular India many hundred million years ago gave rise to Deccan Traps. Basaltic lava flowed out of fissures covering a vast area of about 10 lakh sq km. Basalts are rich in titanium, magnetite, aluminium and magnesium.This is fertile with high moisture holding capacity and is popularly known as 'regur' or black cotton soil.

v. Tertiary and Mesozoic sedimentary rocks- Rocks of extra peninsular (plains and Himalayas) India have given rise to soils with high porosity. These soils are generally immature recent and sub recent rocks, result in alluvial soils on weathering. Alluvial fertile soils consist of fine silts and clay.

2. Climate- Climate directly has an effect on the kind of vegetation in an area which in turn will affect the soil formation processes related to root penetration and vegetation cover. The two most important climatic variables, influencing soil formation are temperature and moisture. Moisture conditions affect the development and character of soil directly than any other climatic factors. High moisture availability in a soil promoted the weathering of bed rock and sediments, chemical reactions and plant growth. Precipitation also affect plant growth, which directly influences a soils organic content and fertility. In wet equatorial climates high rainfall will cause leaching of nutrients and a relatively infertile soil. Evaporation is also an important factor. Salt and gypsum deposits from the upward migration of capillary water are more extensive in hot, dry regions than in colder, dry regions. The availability of moisture also has an influence of soil pH and the decomposition of organic matter. Temperature has a direct influence on the weathering of bedrock to produce mineral particles. Rates of bedrock weathering generally increase with highest temperatures. Chemical activity increases and decreases directly with temperature, given equal availability of moisture.

Climate affects the soil formation directly and indirectly. Directly, climate affects the soil formation by supplying water and heat to react with parent material. Indirectly, it determines the fauna and flora activities which furnish a source of energy in the form of organic matter. This energy acts on the rocks and minerals in the form of acids, and salts are released. The indirect effects of climate on soil formation are most clearly seen in the relationship of soils to vegetation.

3. Organism- All living organisms play an active role in the soil formation processes. Organisms including fungi, bacteria, animals, humans, and vegetations are the major determinants and they impact on the physical and chemical environments of the soils. Micro-organisms encourage acidic conditions which change the soils chemistry and eventually determine the kind of soil formation process that occur. Microbial activities also decompose organic matter and recycle them in the soil. Larger animals including burrowing animals and earthworms mix the soil and alter its physical characteristics. Man's activities have as well made tremendous changes to the natural soils. Through cultivation, construction, and addition of fertilizer and lime has altered the physical and chemical properties of the soil.

4. Relief or Topography- Topography generally modifies the development of soil on a local or regional scale, the slope of the land, its relief, and its aspect all influence soil development. Soils developing on moderate to gentle slopes are often better drained than soils found at the bottom of valleys. Steep topographic gradients inhibit the development of soil because of erosion. The topography refers to the differences in elevation of the land surface on a broad scale. The prominent types of topography designations, as given in FAO Guidelines (1990), shown in Table 4.1.

Table 4.1: Prominent types of topography designations

Land surface	With slopes (in %)
Flat to Almost flat	0-2
Gently undulating	2-5
Undulating	5-10
Rolling	10-15
Hilly	15-30
Steeply dissect	>30% with moderate range of elevation (<300m)
Mountainous	>30% with moderate range of elevation (>300m)

5. Time- Soil formation takes several hundreds to thousands of years to undergo significant changes. Soils can take many years to form. Younger soils have some characteristics from their parent material, but as they age, the addition of organic matter, exposure to moisture and other environmental factors may change its features. With time, they settle and are buried deeper below the surface, taking time to transform. Eventually they may change from one soil type to another. Most of the soils of the world have taken more than 10,000 years to form the current state of soils.

Soil forming processes

The pedogenic processes are extremely complex and dynamic involving many chemical and biological reactions, and usually operate simultaneously in a given area. One process may counteract another or two different processes may work simultaneously to achieve the same result. The ultimate result of soil formation is profile development.

A. Fundamental soil forming processes

i. **Humification-** Humification is the process of decomposition of organic matter and synthesis of new organic substances called humus. It is the process of transformation of raw organic matter into formation of surface humus layer, called Ao- horizon. The percolating water passing through this layer dissolves certain organic acids and affects the development of the lower A-horizon and the B- horizon.

ii. **Eluviation and illuaviation-** Eluviation means washing out. It is the process of removal of constituents by percolation from upper layers to lower layers. This layer of loss is called eluvial and designated as the A-horizon. The eluviated producers move down and become deposited in the lower horizon which is termed as the illuvial or B-horizon. The eluviation produces textural differences. The process of illuviation leads to the textural contrast between A2 and B1 horizon.

B. Specific soil forming processes

i. **Podsolisation-** It is a type of eluviation in which humus become mobile, leach out from upper horizons and become deposited in the lower horizons. This process is favoured by cool and wet climate. It requires high content of organic matter and low alkali in the parent material. The process increases the proportion of silica, sesquioxide in A-horizons and accumulation of clay, iron and aluminum in B-horizons.

ii. **Laterisation-** In this process, silica is removed while iron and alumina remain behind in the upper layers. It is favoured by rapid decomposition of parent rocks under climates with high temperature and sufficient moisture for intense leaching such as found in the tropics. The soil formed in this process is acidic in nature.

iii. **Clacification-** It is the process of precipitation and accumulation of calcium carbonate ($CaCO_3$) in some part of the profile. The accumulation of $CaCO_3$ may result in the development of a calcic horizon. Calcium is readily soluble in acid soil water and/or when CO_2 concentration is high in root zone. This process is favoured by scanty rainfall and alkali in parent material.

iv. **Gleization-** The term gleiis is of Russian origin, which means blue, grey or green clay. The gleization is a process of soil formation resulting in the development of a glei (or gley horizon) in the lower part of the soil profile above the parent material due to poor drainage condition (1ack of oxygen) and where waterlogged conditions prevail. Under such condition, iron compounds are reduced to soluble ferrous forms. This is responsible for the production of typical bluish to grayish horizons with mottling of yellow and reddish brown colours.

v. **Salinization-** Salinization is the process of accumulation of salts, such as sulphates and chlorides of calcium, magnesium, sodium and potassium, in soils in the form of salty (salic) horizons. It is quite common in arid and semi arid regions. It may also take place through capillary rise of saline groundwater and by inundation with seawater in marine and coastal soils. Salt accumulation may also result from irrigation or seepage in area of impeded drainage.

vi. **Desalinization-** It is the removal by leaching of excess soluble salts from horizons or soil profile by ponding water and improving the drainage conditions by installing artificial drainage network.

vii. **Solonization (Alkalization)-** The process involves the accumulation of sodium ions on the exchange complex of the clay, resulting in the formation of sodic soils (solonetz).

viii. **Solidization (Dealkalization)-** The process refers to the removal of Na+ from the exchange sites. This process involves dispersion of clay. Dispersion occurs when Na+ ions becomes hydrated. Much of the dispersion can be eliminated if Ca+ and Mg++ ions are concentrated in the water, which is used to leach the solonetz. These Ca and Mg ion can replace the Na on exchange complex, and the salts of sodium are leached out.

ix. **Pedoturbation-** Another process that may be operative in soils is pedoturbation. It is the process of mixing of the soil e.g. argillipedoturbation is observed in deep black soils.

References

https://en.wikipedia.org/wiki/Pedogenesis

https://www.eartheclipse.com/environment/process-and-factors-of-soil formation.html

Jenny, H. 1941. Factors of Soil Formation, A System of Quantitative Pedology. McGraw- Hill, New York.

Lesson 3. Soil Forming Processes. http://ecoursesonline.iasri.res.in/mod/page/view.php?id =1512

NCERT Geography, Indian Geography by Kullar

5

Saline and Sodic Soils and Their Reclamation and Management

Swati Shabnam, Kerobim Lakra and Soumitra Sankar Das

Introduction

Soils are formed by weathering of rocks and minerals and all soils contain some amount of soluble salts. Many of these act as a source of essential nutrients for the healthy growth of plants. However, when quantity and quality of salts in the soil near rhizosphere exceeds a particular value, growth, yield and/or quality of most crops is adversely affected. Such a soil is called salt-affected. The degree of adverse effects depends upon the type and quantity of salts, crop and its variety, stage of growth, cultural practices and environmental factors viz. temperature, relative humidity, and rainfall etc. Development of salinity and waterlogging is a serious problem in arid and semi-arid regions of the world and threatening the sustainability of irrigated agriculture.

Soil degradation resulting from salinity and sodicity is a major environmental threat to soil fertility and agricultural productivity in arid and semiarid regions of the world. Saline–sodic soils are degraded due to the simultaneous effect of salinity and sodicity. This causes loss of soil physical structure by clay swelling, and dispersion because of high Na^+ concentrations in the soil solution or at the exchange phase. Apart from these physicochemical effects, this leads to biological properties, such as microbial respiration and biomass, becoming worse. On the other hand, salinity affects plant growth by creating osmotic imbalances and specific ion toxicities, which pose limitations to morphological, histological, chemical, biochemical, and metabolic processes. In turn, these reduce stomatal opening and photosynthetic rate, leading to decreased plant growth, crop yield, and quality.

Soil Salinity

The distinguishing characteristic of saline soils from the agricultural standpoint, is that they contain sufficient neutral soluble salts to adversely affect the growth of most crop plants. For purposes of definition, saline soils are those which have an electrical conductivity of the saturation soil extract of more than 4 dS/m at 25°C (Richards 1954). This value is generally used in the world over although the terminology committee of the Soil Science Society of America has lowered the boundary between saline and nonsaline soils to 2 dS/m in the saturation extract. Soluble salts most commonly present are the chlorides and sulphates of sodium, calcium and magnesium. Nitrates may be present in appreciable quantities only rarely. Sodium and chloride are by far the most dominant ions, particularly in highly saline soils, although calcium and magnesium are usually present in sufficient quantities to meet the nutritional needs of crops. Many saline soils contain appreciable quantities of gypsum ($CaSO_4$, $2H_2O$) in the profile. Soluble carbonates are always absent. The pH value of the saturated soil paste is always less than 8.2 and more often near neutrality (Abrol *et al.* 1980).

Salt affected soils differ from arable soils with respect to two important properties, namely, the soluble salts and the soil reaction. A build-up of soluble salts in the soil may influence its behaviour for crop production through changes in the proportions of exchangeable cations, soil reaction, physical properties and the effects of osmotic and specific ion toxicity. Salt-related properties of soils are subject to rapid change. Therefore, to facilitate discussion on soil management and the influence of the two common kinds of salts (neutrals and alkali salts) on soil properties and plant growth, salt affected soils are broadly grouped as either saline or alkali soils (Szabolcs 1979; Abrol and Bhumbla 1979).

Soil sodicity

The chief characteristic of sodic soils from the agricultural stand point is that they contain sufficient exchangeable sodium to adversely affect the growth of most crop plants. For the purpose of definition, sodic soils are those which have an exchangeable sodium percentage (ESP) of more than 15. Excess exchangeable sodium has an adverse effect on the physical and nutritional properties of the soil, with consequent reduction in crop growth, significantly or entirely. The soils lack appreciable quantities of neutral soluble salts but contain measurable to appreciable quantities of salts capable of alkaline hydrolysis, e.g. sodium carbonate. The electrical conductivity of saturation soil extracts are, therefore, likely to be variable but are often less than 4 dS/m at 25 °C. The pH of saturated soil pastes is 8.2 or more and in extreme cases may be above 10.5. Dispersed and dissolved organic matter present in the soil solution of highly sodic soils may be deposited on the soil surface by evaporation causing a dark surface which is why these soils have also been termed as black sodic soils.

Causes of salinity development

There may be a number of causes of soil salinity development such as (1) inherent soil salinity (2) seawater intrusion to coastal areas; (3) uses of brackish/saline water in farming (4) restricted drainage developed into a high water table; (5) low rainfall; (6) high evapotranspiration.

Natural or inherent soil salinity (due to parent material)

Salts are a natural component in soils and water. The ions responsible for salination are: Na^+, K^+, Ca^{2+}, Mg^{2+} and Cl^-. As the Na^+ (sodium) predominates, soils can become sodic. Sodic soils present particular challenges because they tend to have very poor structure which limits or prevents water infiltration and drainage.

Over long periods of time, as soil minerals weather and release salts, these salts are flushed or leached out of the soil by drainage water in areas with sufficient precipitation. In addition to mineral weathering, salts are also deposited via dust and precipitation. In dry regions salts may accumulate, leading to naturally saline soils. This is the case, for example, in large parts of Australia. Human practices can increase the salinity of soils by the addition of salts in irrigation water. Proper irrigation management can prevent salt accumulation by providing adequate drainage water to leach added salts from the soil. Disrupting drainage patterns that provide leaching can also result in salt accumulations. An example of this occurred in Egypt in 1970 when the Aswan High Dam was built. The change in the level of ground water before the construction had enabled soil erosion, which led to high concentration of salts in the water table. After the construction, the continuous high level of the water table led to the salination of the arable land.

Dry land salinity

Arid zones receive inadequate and irregular precipitation to accomplish leaching of salts originally present in the soil profile. Normally when the precipitation is more than 1000 mm per annum, salinity should not develop. This is not the case in arid zones; therefore, salts accumulate in soils. Salt buildup in concentrations detrimental to plant growth is a constant threat in irrigated crop production. In arid and semiarid regions, evapotranspiration is higher than the total annual rainfall. Therefore, rainfall contributes insignificantly to groundwater recharge, and hence there is a general shortage of fresh quality water to offset the total agriculture water demand in these countries. The shortage of fresh water necessitates the use of marginal quality ground water, such as brackish and saline, for irrigated agriculture. This is highly demanded in water-scarce regions. The improper use of saline/brackish water in irrigated agriculture often introduces salinity and sodicity problems and the soil if not properly managed can reach a condition where it cannot be exploited to its full production capacity. Under such conditions, irrigated agriculture has faced the challenge of sustaining its productivity for centuries, particularly soil and water salinity, poor irrigation, and drainage management continue to plague agriculture especially in arid and semi arid regions (Tanji, 1996).

Salinity in drylands can occur when the water table is between two and three metres from the surface of the soil. The salts from the groundwater are raised by capillary action to the surface of the soil. This occurs when groundwater is saline (which is true in many areas), and is favored by land use practices allowing more rainwater to enter the aquifer than it could accommodate. For example, the clearing of trees for agriculture is a major reason for dryland salinity in some areas, since deep rooting of trees has been replaced by shallow rooting of annual crops.

Secondary salinization, or salinity due to irrigation

Salinity from irrigation can occur over time wherever irrigation occurs, since almost all water (even natural rainfall) contains some dissolved salts. When the plants use the water, the salts are left behind in the soil and eventually begin to accumulate. Since soil salinity makes it more difficult for plants to absorb soil moisture, these salts must be leached out of the plant root zone by applying additional water. This water in excess of plant needs is called the leaching fraction. Salination from irrigation water is also greatly increased by poor drainage and use of saline water for irrigating agricultural crops.

Area under saline and sodic soils

Salt-affected soils occupy an estimated 952.2 million ha of land in the world that constitutes nearly 7% of the total land area and 33% of the potential arable land (Dudal and Purnell 1986). In India, the salt affected soils account for 6.727 million ha i.e. 2.1 % of geographical area of the country. These soils are mostly found in the states of Uttar Pradesh, Haryana, Punjab, Gujrat, Madhya Pradesh, Bihar and Andhra Pradesh. In the last 25 years 1.1 million ha of alkali soils have been reclaimed in the states of Haryana, Punjab and Uttar Pradesh. These have contributed to the additional food grain production of 10 million tonnes annually.

Remote sensing techniques and GIS tools have proven to provide useful information on monitoring, delineating and mapping salt affected soils (Arora 2013). Remote sensing based identification of salt affected soils based on differences in spectral signatures using multispectral and hyperspectral data, is the most common approach (Sharma and Mondal 2006; Singh *et al.* 2017).

National Remote Sensing Agency (NRSA), Hyderabad in association with other National and State level organizations like Central Soil Salinity Research Institute, Karnal; National Bureau of Soil Survey & Land Use Planning, Nagpur; All India Soil Survey & Land Use, Delhi; and State Government Agencies conducted survey and used remote sensing data to prepare the maps of salt affected soils of India in 1996. The Landsat satellite images were used in mapping salt affected soils at 1:250,000 scale. Satellite images were interpreted for broad categorization of different types of salt-affected soils, sample areas for field verification (ground truthing) were identified and surveyed for soil sampling and characterization. The salt affected soils were classified according to norms for pH, electrical conductivity (EC) and exchangeable sodium percentage (ESP). The state wise extent of salt affected soils in India is given in Table 1. It shows that maximum area of salt affected soils occur in Gujarat followed by Uttar Pradesh and Maharashtra which account for about 62.4 per cent of the total. Due to the limitation of small scale some very small and isolated patches of salt affected soils occurring in the states of Delhi and Himachal Pradesh could not be detected. Out of the total 6.727 million ha of salt affected soils, 2.956 million ha are saline and the rest 3.771 million ha are sodic. Out of the total 2.347 million ha salt affected soils in the Indo-Gangetic Plains, 0.56 million ha are saline and 1.787 million ha are sodic.

Characteristic of salt affected soils

Electrical conductivity of the soil saturation extract (ECe) is the standard measure of salinity. Richards (1954) has described general relationship of ECe and plant growth.

Measurement of soil salinity

The standard unit of conductance is siemens and when expressed per unit of distance, the standard unit of conductivity is siemens per metre. The conductivity of most saturation paste extracts is only a fraction of a siemens per metre. For convenience, therefore, conductivities of soil extracts are expressed in deci Siemens (mS) per metre at 25°C. EC measurements are quick and sufficiently accurate for most purposes.

To determine EC the solution is placed between two electrodes of constant geometry and constant distance of separation. When an electrical potential is imposed, the amount of current varies directly with the total concentration of dissolved salts. At constant potential, the current is inversely proportional to the solution's resistance and can be measured with a resistance bridge. Conductance is the reciprocal of resistance and has the unit Siemens (formerly, mhos).

The measured conductance is the result of the solution's salt concentration and the electrode geometry. The effects of electrode geometry are embodied in the cell constant and this is related to the distance between electrodes divided by their effective cross sectional area. The cell constant is commonly obtained by calibration with KCl solutions of known concentration.

Effect of salt on crop growth

Excess soil salinity causes poor and spotty stands of crops, uneven and stunted growth and poor yields, the extent depending on the degree of salinity. The primary effect of excess salinity is that it renders less water available to plants although some is still present in the root zone. This is because the osmotic pressure of the soil solution increases as the salt concentration increases. Apart from the osmotic effect of salts in the soil solution, excessive concentration and absorption of individual ions may prove toxic to the plants and/or may retard the absorption of other essential plant nutrients.

There is no critical point of salinity where plants fail to grow. As the salinity increases growth decreases until plants become chlorotic and die. Plants differ widely in their ability to tolerate salts in the soil. Salt tolerancc ratings of plants are based on yield reduction on salt-affected soils when compared with yields on similar non-saline soils. Soil salinity classes generally recognized are given in Table 5.1.

Table 5.1: Soil salinity classes and crop growth

Class	ECe (ds m^{-1})	Plant growth
0 Non saline	0–2	Salinity effects mostly negligible
1 Very slightly saline	2–4	Yields of very sensitive crops may be restricted
2 Slightly saline	4–8	Yields of many crops restricted
3 Moderately saline	8–16	Only tolerant crops yield satisfactory
4 Strongly saline	>16	Only a few very salt-tolerant crops yield satisfactory

Reclamation of saline and sodic soils

Therefore, a successful reclamation of saline–sodic soils is of great importance for crop productivity, and a key factor in addressing the world's food security challenges. Reclamation of saline–sodic soils requires the removal of sodium from the soil exchange sites by divalent cations (preferably Ca^{2+}) to promote soil flocculation. The most common amendment source can be calcium, which provides soluble calcium within the soil. The replaced Na^{+} is removed either below root zone or out of the soil profile by leaching water. Gypsum is the most commonly applied product for the reclamation of saline–sodic soils and can improve physical and chemical soil properties primarily by maintaining a favorable electrolyte concentration in soil solution. As adsorbed Na^{+} on exchangeable sites of clay particles are considered to be responsible for soil dispersion, gypsum can prevent it by maintaining high Ca/Na ratios, and thus promoting clay flocculation and structure stability. On the other hand, organic amendments can increase dissolution of native calcite ($CaCO_3$) minerals via increased formation of carbonic acid in the soil profile, thus dissolution could release Ca^{2+} in the soil solution to facilitate the removal of Na^+ from cation exchange sites. Therefore, the addition of organic matter also represents an alternative for reclaiming a variety of degraded soils, including salt-affected soils. In fact, significant improvements have been reported regarding soil physical, chemical, and biological characteristics with the use of organic matter.

Removal of salts

The amount of crop yield reduction depends on such factors as crop growth, the salt content of the soil, climatic conditions, etc. In extreme cases where the concentration of salts in the root zone is very high, crop growth may be entirely prevented. To improve crop growth in such soils the excess salts must be removed from the root zone. The term reclamation of saline soils refers to the methods used to remove soluble salts from the root zone. Methods commonly adopted or proposed to accomplish this include the following:

Scraping: Removing the salts that have accumulated on the soil surface by mechanical means has had only a limited success although many farmers have resorted to this procedure. Although this method might temporarily improve crop growth, the ultimate disposal of salts still poses a major problem.

Flushing: Washing away the surface accumulated salts by flushing water over the surface is sometimes used to desalinize soils having surface salt crusts. Because the amount of salts that can be flushed from a soil is rather small, this method does not have much practical significance.

Leaching: This is by far the most effective procedure for removing salts from the root zone of soils. Leaching is most often accomplished by ponding fresh water on the soil surface and allowing it to infiltrate. Leaching is effective when the salty drainage water is discharged through subsurface drains that carry the leached salts out of the area under reclamation. Leaching may reduce salinity levels in the absence of artificial drains when there is sufficient natural drainage, i.e. the ponded water drains without raising the water table. Leaching should preferably be done when the soil moisture content is low and the groundwater table is deep. Leaching during the summer months is, as a rule, less effective because large quantities of water are lost by evaporation. The actual choice will however depend on the availability of water and other considerations. In some parts of India for example, leaching is best accomplished during the summer months because this is the time when the water table is deepest and the soil is dry. This is also the only time when large quantities of fresh water can be diverted for reclamation purposes.

Quantity of irrigation water for leaching

It is important to have a reliable estimate of the quantity of water required to accomplish salt leaching. The initial salt content of the soil, desired level of soil salinity after leaching, depth to which reclamation is desired and soil characteristics are major factors that determine the amount of water needed for reclamation. A useful rule of thumb is that a unit depth of water will remove nearly 80 percent of salts from a unit soil depth. Thus 30 cm water passing through the soil will remove approximately 80 percent of the salts present in the upper 30 cm of soil. Similarly, to reduce the salt content of the surface 60 cm of soil to about 20 percent of the original value would require the passage of about 60 cm of water through the soil. For more reliable estimates, however, it is desirable to conduct salt leaching tests on a limited area and prepare leaching curves. Leaching curves relate the ratio of actual salt content to initial salt content in the soil (Sa/Sb) to the depth of leaching water per unit depth of soil (Dw/Ds).

Water application method

Results from several laboratory experiments and some field trials have shown that the quantity of salts removed per unit quantity of water leached can be increased appreciably by leaching at soil moisture contents of less than saturation, i.e. under unsaturated conditions. In the field unsaturated conditions during leaching were obtained by adopting intermittent ponding or by intermittent sprinkling at rates less than the infiltration rate of the soil. Nielsen *et al.* (1966) for example, showed that 25 cm of sprinkled water reduced the salinity of the upper 60 cm of soil to the same degree as 75 cm of ponded water. In both irrigation methods, at the end of irrigation, upper parts of the soil profiles have low concentration of salts and these will depend on the salt concentration of the applied irrigation water. The salt in the profile increases to a maximum value close to the wetting front and drops to its initial value below the wetting depth. Because of a slower wetting rate under sprinkling, the zone of complete leaching at the end of irrigation extends more deeply into the profile than under flood irrigation. When the soil is subjected to evaporation, water carrying salts moves simultaneously in the upward and downward directions. Thus some salts continue to move down with the redistributed water and, at the same time, salts near the surface move towards the soil surface where they accumulate. The amount of salts which move to the surface depend on the amount of salts present in the upper soil layers from where the water can flow upwards. Thus, only a small fraction of salts move up during evaporation from the soil previously irrigated by sprinklers. In flooded soils, on the other hand, more salts move upward and accumulate in the soil surface.

Amendments

Whether an amendment (e.g. gypsum) is necessary or not for the reclamation of saltaffected soils is a matter of practical importance. Saline soils are dominated by neutral soluble salts and at high salinities sodium chloride is most often the dominant salt although calcium and magnesium are present in sufficient amounts to meet the plant growth needs. Since sodium chloride is most often the dominant soluble salt, the SAR of the soil solution of saline soils is also high. When a soil solution is diluted by a factor X, the reduced ratio

$$Na^+ / \frac{\sqrt{Ca^{++} + Mg^{++}}}{2}$$

will decrease by a factor . This implies that desodication will always accompany desalinization. A favourable calcium to sodium ratio of the irrigation water and any supply of inherent calcium from the soil is likely to further accelerate the desodication process.

Gupta *et al.* (2016) found that the combined application of gypsum and organic amendments in sodic soils improved soil properties, resulting in decreased soil bulk density, electrical conductivity (EC), and exchangeable sodium percentage (ESP), but also in increased soil biological activity. Humic substances are widely recognized as key components of soil fertility because they control the physicochemical and biological properties of the rhizosphere, perform various functions in the soil and regulate plant growth. In fact, humic substances improve soil properties, such as aggregation, aeration, and water holding capacity. Furthermore, Ciarkowska *et al.* (2017) reported that humic substances applied to the soil contributed substantially to improve soil quality status, resulting in increased organic carbon, enzymatic activity, root biomass, and yield.

Biochar is another soil amendment that can improve soil physical conditions. It is more recalcitrant than humic substances and less likely to directly support microbial growth, but it can significantly promote soil biological activity due to the release of temporary labile pyrolysis products; however, this effect could be very temporary. Salt-affected soils can benefit from biochar application by increased contents of soil organic carbon and nutrients (K^+, Ca^{2+}, Mg^{2+}, Zn, Mn), increased surface area, enhanced physical properties by balancing water content and air porosity, increased retention of polyvalent cations, and replacement of Na^+ from exchange sites by providing Ca^{2+} in soil solution. In this sense, the addition of organic amendments in combination with gypsum has proven successful in reducing adverse effects on soil properties associated with a high content of sodium.

References

Abrol, I.P., Chhabra, R. and Gupta R. K. 1980. A fresh look at the diagnostic criteria for sodic soils. In: International symposium on salt affected soils. Central Soil Salinity Research Institute, Karnal, February 18-21, pp. 142-147.

Abrol, I.P. and Bhumbla, D.R. 1979. Crop responses to differential gypsum applications in a highly sodic soil and the tolerance of several crops to exchangeable sodium under field condition. *Soil Science,* 127: 79-85.

Arora, S. 2013. Remote sensing and Geographic Information System approach for mapping and reclamation of SaltAffected Soils. In: Participatory approaches to enhance farm productivity of salt affected soils. Singh et al (eds). IRRI-CSSRI, RRS, Lucknow, pp. 9-32.

Ciarkowska, K., Solek-Podwika, K., Filipek-Mazur, B. and Tabak, M. 2017. Comparative effects of lignite-derived humic acids and FYM on soil properties and vegetable yield. *Geoderma,* 303: 85–92.

Gupta, M., Srivastava, P.K., Niranjan, A. and Tewari, S.K. 2016. Use of a bioaugmented organic soil amendment in combination with gypsum for Withania somnifera growth on sodic soil. *Pedosphere,* 26: 299–309.

Nielsen, D.R., Biggar, J.W., Luthin, J.N. 1966. Desalinization of soils under controlled unsaturated conditions. In Proceedings, International Commission on Irrigation and Drainage, 6th Congress, New Delhi, India.

Richards, L. A. 1954. Diagnosis and Improvement of Saline and Alkali Soils. U.S. Department of Agriculture. Handbook 60. U.S. Gov. Printing Office, Washington, DC.

Sharma, R.C. and Mondal, A.K. 2006. Mapping of soil salinity and sodicity using digital image analysis and GIS in irrigated lands of the Indo-Gangetic Plain. *Agropedology,* 16: 71–76.

Singh, R.P., Setia, R., Verma, V.K., Arora, S., Kumar, P. and Pateriya, B. 2017. Satellite remote sensing of saltaffected soils: Potential and limitations. *Journal of Soil and Water Conservation,* 16(2): 97-107.

Szabolcs, I. 1979. Review of research on salt affected soils. Natural Resource Research, IS, UNESCO, Paris.

Tanji, K. K. 1996. Nature and extent of agricultural salinity. In Agricultural Salinity Assessment and Management Manual, K. K. Tanji (ed.). ACSE Manuals and Reports on Engineering Practice No.71, New York, pp.1–17.

6

Acid Soils and Acid Sulphate Soils and Their Reclamation and Management

Asisan Minz and Rakesh Kumar

Introduction

The optimum soil pH for plant production is one that is slightly acidic, at this pH soil microorganisms are most active and plant nutrients are readily available. At extremes of high (alkaline) and low (acidic) pH this delicate balance is disturbed and plant nutrients that were in adequate supply can become either deficient or toxic to plant growth. Some essential nutrients such as phosphorous, calcium, magnesium and molybdenum become unavailable if the soil pH becomes too acidic. Acidic conditions will result in a lowering of plant production in farming systems. This will result in reduced profitability and an increased reliance on fertilizers to sustain any form of productive agriculture. Correcting soil pH to a more favorable pH range will increase the availability of essential nutrients.

Soil acidity is a serious constraint to crop production and ecosystem health. Acidification can occur naturally via the formation of carbonic acid ($CO_2 + H_2O \longrightarrow H_2CO_3 \longrightarrow HCO_3^- + H^+$) and subsequent leaching of basic cations or through man-made processes. Soil acidity is an important chemical property of soil having profound influence on soil biological and chemical properties and plant nutrient availability. Soil acidification is the accumulation of acid in the soil. It is a natural process, in natural ecosystems, operates over many thousands of years, but today faulty agricultural practices and management, can accelerate this process. The rate of soil acidification depends on the rate of addition of acidity to the soil and is broadly related to soil type, rainfall, land use, etc. Before discussing soil acidity, it is necessary to understand about the acids, bases and soil reactions (soil pH).

Starting simply, 'acids' are the substances that produce hydrogen ions in water or give up a proton. So we can define an acid as a substance that when added to water increases the number of H^+ ions in water, and 'bases' are the substances which release OH^- ions in water or accept a proton. It can be defined as a substance that when added to water increases the number of OH^- ions in water. A common example of an acid is sulphuric acid (H_2SO_4), when it is diluted with water the acid gives a proton to form a new acid H_3O^+ (hydronium ion) and a new base HSO_4^-.

$$H_2SO_4 + H_2O \longleftrightarrow H_3O^+ + HSO_4^-$$

$$H_2SO_4 + HOH \longleftrightarrow 3H^+ + SO_4^{2-} + OH^-$$

It is a strong acid because it contains more hydrogen ions (H^+) than hydroxyl ions (OH^-).

Sodium hydroxide (NaOH) is an example of a base, when NaOH is mixed with water; it ionizes, and forms excess hydroxyl ions (OH^-) than hydrogen ions (H^+). This results in a strong basic solution:

$$NaOH + HOH \leftrightarrow Na^+ + H^+ + 2OH^-$$

Soil pH

The term pH stands for the potential (p) of hydrogen ions (H^+) in water, and soil pH is a characteristic that describes the relative acidity or alkalinity of the soil solution. Soil pH or soil reaction, is defined as the negative logarithm of the concentration of hydrogen ion (H^+) activity in soil solution under soil-water system. Mathematically,

$$pH = \log 1/[H^+] = -\log[H^+]$$

It indicates whether the soil is acidic, alkaline or neutral in reaction. The acidity or alkalinity of a solution can be expressed on the scale of acidity or alkalinity, which is known as pH scale and the unit of this scale, is called pH value. The term pH was first coined and used by Danish Scientist, Sorensen in 1909. This pH scale runs from 0 to 14. A pH of 7.0 is neutral (concentration of H^+ ion is equal to that of OH^- ions), values below 7.0 are acidic (H^+ ions predominate), and those above 7.0 are basic / alkaline (OH^- ions predominate). In practical terms, soils between pH 6.5 and 7.5 are considered neutral. Soils in the range 5.6 to 6.0 are moderately acidic and below 5.5 strongly acidic. Soil pH is the measurement of the concentration of hydrogen ions in the soil solution. When the concentration of hydrogen ions in the soil water solution is high, the pH is low and the lower the pH value, higher is the acidity of the soil.

Because pH is expressed on a logarithmic scale, each change of one pH unit actually represents a 10-fold increase in soil acidity or alkalinity, which means the H^+ ion concentration, has a 10 fold change between each whole pH number. For example, a soil with pH of 5 is 10 times more acidic than a soil with a pH of 6 and is 100 (10^2) times more acidic than a soil with a pH of 7.0, shown in table 6.1.

Table 6.1: Degree of soil acidity and the pH range

Reaction	pH	Reaction	pH
Ultra acid	< 3.5	Moderately acid	5.6 – 6.0
Extremely acid	3.5 – 4.4	Slightly acid	6.1 – 6.5
Very strongly acid	4.5 – 5.0	Neutral	6.6 – 7.3
Strongly acid	5.1 – 5.5	-----	-----

Since crop growth is affected much both under low (strongly acidic) as well as very high pH (alkaline) conditions, appropriate reclamation measures becomes essential. It affects the availability of minerals and nutrients to plants as well as many soil processes.

Soil acidity

Acid soil is a base unsaturated soil which has got enough of adsorbed exchangeable hydrogen ions (H^+) instead of OH^- ions so soil tends to have a pH of lower than 7.0. Soil acidity occurs when there is a build up of acid forming elements in the soil. Soil acidity involves both intensity and quantity aspects. The intensity aspect is measured by the measurements of H^+ ion activity, expressed as pH with the help of pH meter. The quantity aspect is associated with, directly or indirectly, by the quantity of liming substances required to neutralize acidity of soil upto desired level for proper crop growth. Soil acidity is a major limitation to the productivity of our soils as it reduces the availability of major soil nutrients (N, P & K) and the uptake and efficiency of applied nutrients in manures or fertilizers.

The range between 5.5 and 7.5 is favourable for two reasons. It allows sufficient microorganisms to break down organic matter. It is also the best range for nutrient availability.

Characteristics of acid soils

- These soils are high in exchangeable hydrogen (H^+) and aluminium (Al^{3+}) ions, with a low pH (< 5.5) and respond to lime application.
- Kaolinitic and illitic types of clay minerals are dominant in acid soils.
- The acid soils have low cation exchange capacities and hence, has high base unsaturation percentage.
- The acid soils contain excessive amounts of soluble aluminum, iron and manganese in toxic concentration, which is very harmful for crop growth.

- The acidic soils are generally deficient in available phosphorus and also available molybdenum because it reacts with the silicates, iron and aluminum compounds. Phosphate, if present as inorganic orthophosphate ions ($H_2PO_4^-$), precipitated as insoluble hydroxyl phosphates, because acid soils have high phosphate fixation capacity.
- Deficiency of bases like Ca^{2+} and Mg^{2+} are found in acid soils.
- The population of bacteria and actinomycetes are lower and cannot function properly in acidic soils; whereas fungi dominate as soil microorganisms.

Sources of soil acidity

The rate of soil acidification process is generally very slow under natural conditions. However, in recent decades, various anthropogenic activities and productive agriculture have accelerated soil acidification to a great extent. Soil acidifies because the concentration of hydrogen ions in the soil increases.

Leaching of bases due to heavy rainfall

Acid soils are common in humid regions and high rainfall areas, where rainfall is greater than evapo-transpiration, and that leads to leaching, rainfall act as a substantial, where, high rainfall is capable enough to leach out appreciable amounts of soluble and exchangeable bases and salts (e.g. Ca^{++}, Mg^{++} , Na^+ and K^+) down the soil profile by excess rains, leaving behind more stable and insoluble compounds, rich in iron and aluminum oxides or hydroxides, resulting in soils that are acidic in nature. Its (Fe and Al) oxides and hydroxides react with water (H_2O) and release hydrogen (H^+) ions in soil solution and soil becomes acidic. When all the soluble bases are lost, the H^+ ions of the carbonic acid and other acids developed in the soil replace the basic cations of the colloidal complex and increases acidity. As indicated below, CO_2 combines with H_2O to form carbonic acid which is responsible for the leaching losses of calcium and magnesium from soil surface layers.

$$CO_2 + H_2O = H_2CO_3 \text{ (Carbonic acid)}$$

$$2H_2CO_3 + CaCO_3 = Ca(HCO_3)_2 \downarrow + 2H^+$$

(Leachable)

Acidic parent material

Soils become acidic due to the parent material being acidic (to be acidic due to leaching of cations), such as granite and that can also contribute to the soil acidification upto some extent.

Application of acid forming fertilizers

The most important cause of soil acidification on agricultural land, however, are the application of ammonium-based fertilizers, urea and elemental S fertilizer. Acid-forming fertilizers are defined as those that lower rhizosphere pH after being absorbed by plants.

Soil management practices, such as regular application of nitrogenous fertilizers, can also acidify soils. Specifically, fertilizers that produce ammonium (NH_4^+), such as ammonium sulphate [$(NH_4)_2SO_4$], ammonium nitrate (NH_4NO_3), urea and anhydrous ammonia, release hydrogen ions (H^+) through a biological process in which N is oxidized to nitrate (NO_3^-), according to

$$(NH_4)_2SO_4 + 4O_2 \xrightarrow{\text{nitrifying bacteria}} 2NO_3^- + SO_4^{2-} + 2H_2O + 4H^+$$

The acidification caused by the use of ammonium fertilizers are explained by the release of H^+.

$$NH_4^+ + 2O_2 \longleftrightarrow NO_3^- + H_2O + 2H^+$$

If the nitrate (NO_3^-) is taken up by the crop, there is no net acidification because the NO_3^- takes up protons with it. Acidification only occurs when NH_4^+ is nitrified and the NO_3^- leached.

When ammonium sulphate [$(NH_4)_2SO_4$] is applied to the soil, this gives ammonium (NH_4^+) ions and acid forming sulphate (SO_4^{2-}) ions. These ammonium (NH_4^+) ions replace calcium (Ca^{2+}) ions from the exchange complex from soils and the calcium sulphate ($CaSO_4$) is formed and finally leached out.

$$(NH_4)_2SO_4 \rightleftharpoons 2NH_4^+ + SO_4^{2-}$$

$$2NH_4^+ + Ca\text{(Clay)} \longrightarrow CaSO_4\downarrow + \text{Clay}(NH_4)_2$$

Leached out

$$\text{(NH4)Clay(NH4)} + 3O_2 \xrightarrow{\text{Nitrification}} \text{(H)Clay(H)} + 2HNO_3$$

Acid soil

Release of H^+ from nitrification and S oxidation

Soil acidity can arise from several processes; some processes are biologically mediated, others are purely chemical.

1.) Nitrification

$$NH_4^+ + 2O_2 \rightleftharpoons NO_3^- + 2H^+ + H_2O$$

In this process, NH_4^+ is converted to NO_3^- by nitrifying bacteria, and H^+ is produced as a by-product.

2.) Oxidation of pyrite (Fe_2S) and elemental S

$$S + 3/2O_2 + H_2O \rightleftharpoons SO_4^{2-} + 2H^+$$

$$2FeS_2 + 7H_2O + 7\ ½\ O_2 \rightleftharpoons SO_4^{2-} + 8H^+ + 2Fe(OH)_3$$

These reactions often occur in strip mine soils where pyrite is abundant or in acid sulphate soils, sometimes called cat clays. When these soils are drained and exposed to oxidation, pH values as low as 3.5 have been observed.

Hydrolysis of Al

H^+ ions directly reduce the pH of the soil solution, while aluminum participates in hydrolysis reactions because it is soluble at low pH. At low pH the concentration of Al^{3+} in the soil solution is high due to dissociation of $Al(OH)_3$ into Al^{3+} and H_2O.

Hydrolysis

$$Al^{3+} + H_2O \rightleftharpoons Al(OH)^{2+} + H^+$$

$$Al(OH)^{2+} + H_2O \rightleftharpoons Al(OH)_2^+ + H^+$$

$$Al(OH)_2^+ + H_2O \rightleftharpoons Al(OH)_3 + H^+$$

$$Al(OH)_3 + H_2O \rightleftharpoons Al(OH)_4^- + H^+$$

All the above products: $Al(OH)^{2+}$, $Al(OH)_2^+$, $Al(OH)_3$, and $Al(OH)_4^-$ are called monomeric hydroxyaluminum species because only one Al atom is involved in those complexes. On later H^+ ions are formed in the soil solution, which tends soil towards acidity.

Trivalent Al^{3+} hydrolyzes easily to form monomeric and polymeric hydroxyl-aluminum complexes and produces H^+.

$$Al^{3+} + 6H_2O \rightleftharpoons Al\ (H_2O)_6^{3+} \rightleftharpoons Al(OH)(H_2O)_5^{2+} + H^+$$

Organic matter decay

Decaying of organic matter to form humus also produces H^+ ions by oxidation of organic matter which is responsible for acidity. The carbon dioxide (CO_2) produced by decaying organic matter reacts with water in the soil to form a weak acid called carbonic acid (H_2CO_3). The amount of carbonic acid present depends on the amount of carbon dioxide (CO_2) in the soil. This is a slow process but it does decrease the soil pH.

Crop production

Harvesting of crops also acidifies soils by removing cations because harvested portions of crops contains more cations (calcium, magnesium, potassium, sodium etc.) than anions (phosphate, sulphate, nitrate, chloride), then there will be a net loss of basic cations from the soil. Increasing crop yields will cause greater amounts of basic material to be removed, which increases soil acidity.

Farming practices

Continuous cultivation of legume crops decreases the pH of agricultural soils and can increase soil acidity. Leguminous species, increase soil acidification in arable cropping systems due to their high absorption of basic cations and the release of H^+ ions by the roots to maintain ionic balance during N_2 fixation. In addition, they state that the amount of H^+ ions released during N_2 fixation is really a function of carbon assimilation and hence depends mainly on the form and amount of amino acids and organic acids synthesized within the plants. Soil acidification is also caused by the release of protons (H^+) during the transformation and cycling of carbon, nitrogen, and sulphur in the soil–plant–animal system.

Acid rain

Acid rain contributes to soil acidity, although the effect is small compared to other sources of acidity. The pH of rainwater is about less than 5.6. It has been known that the high concentrations of HNO_3 and H_2SO_4 in acid rain are due to atmospheric oxidation of NO_x and SO_2 emitted by fossil fuel combustion. The sources of this additional acidity are sulphur dioxide (SO_2) from the combustion of sulphur -containing fuels (coal and oil) and from the smelting of sulfur-containing ores (mostly copper, lead, and zinc) and volcanic activity; oxidation of hydrogen sulfide (H_2S) from pulp mills and swamps; and nitrous and nitric oxides from thunderstorms and emissions from internal combustion engines, and from conversion of soil nitrate.

Beyond the input of acidity, deposition of NH_4^+ and NO_3^- fertilizes ecosystems by providing a source of directly assimilable nitrogen. This source has been blamed as an important contributor to the eutrophication (excess fertilization); a consequence of this eutrophication is the accumulation of algae at the surface of the water bodies, suppressing the supply of O_2 to the deep-water biosphere.

Kinds of soil acidity

There are three major types of acidity namely, active acidity, exchangeable acidity, and residual or reserve acidity.

i. **Active acidity** is also known as actual acidity and it is the measurement of free hydrogen ions (H^+) activity in soil solution. Active acidity can be defined as the acidity developed due to the accumulation of hydrogen (H^+) ions in the soil solution. The magnitude of active acidity is very small, compared to other acidity, because in active acidity the hydrogen ions are in equilibrium with the exchangeable hydrogen ions that are held on the soil's cation exchange complex. That's why implying only a meager amount of lime would be required to neutralize active acidity. This acidity most readily affects plant growth.

ii. **The exchangeable acidity** is the measure of the H^+ and Al^{3+} ions retained or fixed on soil colloid after the active acidity is measured. The exchangeable acidity, can be defined as the acidity develops due to the adsorbed aluminium (Al^{3+}) and hydrogen (H^+) ions, occupied on the soil colloids. When the cation exchange capacity of a soil is high but has a low base saturation, the soil becomes more resistant to pH changes, so their magnitude is very high. As a result, it will require larger additions of lime to neutralize the acidity. The soil is then buffered against pH change.

iii. **Residual or reserve acidity** may be defined as the acidity which remains in the soil after the neutralization of active and exchangeable acidity in reserve form. Aluminum hydroxyl ions *viz.* $Al(OH)^{2+}$ and $Al(OH)_2^+$ and H and Al ions present in non- exchangeable form with organic matter and clay account for the residual acidity in soil. Out of all acidity, residual acidity is least available.

Thus, Total acidity can be defined as the summation of active, Exchangeable and residual acidity.

Total acidity = Active acidity + Exchangeable acidity + Residual acidity

Total acidity is dependent on the initial and final pH values of the soil and on the operational conditions such as stirring time, waiting time between each increment of base.

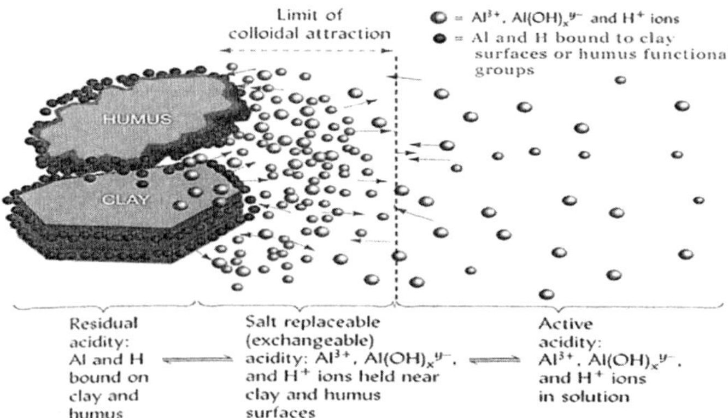

Fig. 6.1: Relationship between reserve, exchange and active acidity

Chemistry of aluminium in soil acidity

Acidity itself is not responsible for restricting crop growth, but the associated chemical changes in acidic soil can restrict the availability of some beneficial plant nutrients and increase the availability of toxic elements. The hydrogen (H^+) ions are directly involved in the soil acidification, while aluminium (Al^{3+}) ions are indirectly by their hydrolysis. The relationship between Al activity and the pH level of acidic soils led to the Al toxicity and it could be a major factor of acid soil infertility.

The solubility of Al is highly pH-dependent and increase in soil pH alleviates its toxicity. As pH decreases, the concentration of Al^{3+} in the soil solution increases due to dissolution of $Al(OH)_3$, because at low pH aluminium is soluble.

$Al(OH)_3 + 3\ H^+\ Al^{3+} + 3\ H_2O$

Soluble Al species can be broadly divided into two groups: monomers and polymers. In acid soils the soluble Al mainly is in monomeric forms. Monomeric Al species are divided into inorganic Al, organic Al and acid soluble Al. Soil Al exists in various chemical forms from soluble ionic species in soil solution to being a basic structural component of the secondary clay minerals such as kaolinite and vermiculite. Aluminium is adsorbed on organic and mineral colloidal surfaces, amorphous and crystalline forms of hydroxy-Al as well as Al polymerized on clay surfaces and between clay inter-layer. In acid soils, an appreciable portion of the cation exchange capacity is satisfied by Al ions *i.e.* soils have a high Al saturation. While these Al ions are referred to as exchangeable Al^{3+}, they are a mixture of monomeric Al ions [Al^{3+}, $Al(OH)^{2+}$ and $Al(OH)_2^{+}$] with an average charge per Al between 2 and 3, decreasing as pH increases. These aluminium hydroxyl ions are adsorbed on soil colloids and act as exchangeable ions. These hydroxyl ions remain in equilibrium with aluminum ions in the soil solution. In the soil solution, these hydrolyze and produce H^+ ions and thus contribute to soil acidity. At different pH levels, Al hydrolyses as,

Stepwise hydrolysis	*Dominant aluminium*	*pH levels*
$Al^{3+} + H_2O$	$Al(OH)^{2+} + H^+$	< 4.6
$Al(OH)^{2+} + H_2O$	$Al(OH)_2^{+} + H^+$	4.6 – 6.5
$Al(OH)_2^{+} + H_2O$	$Al(OH)_3^{0} + H^+$	6.5 – 8.0
$Al(OH)_3^{0} + H_2O$	$Al(OH)_4^{-} + H^+$	8.0 – 11.0

As monomeric Al dominants in strongly acid soils it is likely to be toxic to plant growth.

Effects of Soil Acidity

Soil acidity affects plants in various ways: nutrient availability, toxicity of some metals, activity of beneficial microorganisms and nitrogen fixation by leguminous crops. Productivity of acid soils, occupying about 30% of the total cultivable area of India, is low, mainly due to the impaired environment of acidic soil, which induces deficiency and toxicity of mineral nutrients. Problems associated with soil acidity may be divided into three groups which are as follows:

1. **Toxic effects**
   (a) Acid toxicity
   (b) Nutrient toxicity in acidic soils
2. **Availability of Nutrients**
3. **Microbial activity**

1. Toxic effects

(a) Acid toxicity

Hydrogen ion, itself, is not very important in mineral soils as far as its phytotoxicity is concerned because H^+ does not become phytotoxic until the pH drops below 4.0 but mineral soils seldom become that acid. The higher concentration of hydrogen ion is toxic to plants under lower pH.

(b). Nutrient toxicity in acidic soils

In acidic soils, with low pH levels, metals such as Aluminium, Manganese and Iron might be released into the soil solution at high concentrations which may be toxic to many plants, whereas at high pH, these ions become deficient.

Aluminium toxicity

When soil pH drops, aluminium becomes soluble. A small drop in pH can result in a large increase in soluble aluminium. In acid soils with a high mineral content, the primary factor limiting plant growth is aluminium toxicity. The primary target of aluminium toxicity is the root apex. Exposure to aluminium causes stunting of the primary root and inhibition of lateral root growth. Affected root tips are stubby due to inhibition of cell elongation and cell division. Typically, roots become thick and stubby and there is little development of the all-important fine roots. The resultant inefficient root systems limit yields through poor water and nutrient uptake. Aluminium toxicity restricts the absorption and translocation of some important nutrient elements from soil to plant like phosphorus. Excess aluminium accumulates in the roots and reduces their power of translocation of phosphates from the soil to the vascular system; hence Al^{3+} ions reduce the availability of phosphorus. Orthophosphate ions $H_2PO_4^-$ are precipitated as insoluble hydroxy phosphates. It occurs as follows:

$$Al^{+++} + H_2PO_4^- + 2OH \longleftrightarrow Al\begin{cases} OH \\ OH \\ H_2PO_4 \end{cases} + 2\,H^+$$

Insoluble aluminium hydroxy phosphate

The aluminium toxicity greatly depends on the degree of relationship between the quantity of aluminium and the quantity of bases (such as Ca and Mg), or their accompanying cations.

Manganese toxicity to plants

Excessive manganese is a problem in extremely acid soils (pH <5.0). Manganese becomes more soluble in acid soils and is thus more likely to cause toxicity than deficiencies. Toxicity may also be induced by water logging.

$$MnO_2 + 4H^+ + 2e^- \rightleftharpoons Mn^{2+} + 2H_2O$$

As a result of reduction, the element Mn^{4+} reduces to Mn^{2+} and increases their concentration to a very high and cause toxicity. The toxic level of Mn in soil solution is about 0.5-5ppm and Mn toxicity often affects the above-ground portion of plants such as developing buds and leaf edges more than the roots. Excess manganese accumulation will take place in all tissues and interferes with their proper metabolism.

Iron toxicity

Iron toxicity commonly occurs in a wide range of acid soils, particularly in lowland rice with permanent flooding where anaerobic conditions occur. Under these conditions, Fe^{3+} ions are readily reduced to more soluble Fe^{2+} ions, which are absorbed by rice plant in larger quantities and causes Fe-toxicity. Excess concentration of reduced iron (Fe^{2+}) results in a range of nutrient disorders and results in deficiencies of other macro and micro nutrients. Nutritional disorders associated with iron toxicity have been divided into direct and indirect toxicity.

Direct toxicity is related to the plant's excessive iron absorption. This excessive absorption damages the plant cells. Symptoms appear initially on younger leaves, where the element concentrates in small brown dots. This phenomenon is known as bronzing. Under extreme situation, the leaves may become chlorotic and at advanced toxicity stages necrosis may occur, i.e. the leaves dry and eventually die.

Indirect toxicity results from the limited absorption of several nutrients such as calcium, magnesium, potassium, phosphorous and iron itself, due to iron precipitation on rice root epidermis. The formation of an oxide-hydroxide Fe^{3+} layer on the root blocks nutrient absorption, resulting in multiple nutritional deficiencies. Symptoms of this deficiency include plant atrophy, tillering reduction, orange leaves, and the covering of roots by red layers of iron oxides.

2. Availability of nutrients

Although pH does not directly affect plants, it does affect the availability of different nutrients to plants. As we know nutrients need to be dissolved in the soil solution before they can be accessed by plants. The soil pH changes whether a nutrient is dissolved in the soil solution or forms other less-soluble compounds. The nutrients availability such as nitrogen, phosphorus and potassium is generally reduced as soil pH decreases.

The desirable soil pH range for optimum plant growth varies among crops. Generally, at slightly acidic pH 6.0-7.5 is acceptable for most plants as most nutrients become available in this pH range. The acidity of the soil influences the solubility of minerals, and thus affects the availability of nutrients. In highly acidic soil, aluminum and manganese can become more available and more toxic to plant while, all the major plant nutrients and also the trace element *e.g.*, molybdenum may be unavailable, or only available in insufficient quantities.

Phosphorus deficiency

Phosphorus is particularly sensitive to pH and can become a limiting nutrient in strongly acid soils. Soil acidity can rapidly fix applied P into relatively unavailable forms. P deficiency is difficult to diagnose and severe yield depressions may often occur without showing leaf deficiency symptoms. Deficiency of phosphorous in acidic soils is caused because of aluminum interference with phosphorous.

The availability of soil phosphate (largely $H_2PO_4^-$ and HPO_4^{2-}) decreases significantly under acidic conditions because of the formation of insoluble compounds like, Al and iron phosphate.

$$H_2PO_4^- + Al(OH)_3 + H^+ \rightleftharpoons Al(OH)_2H_2PO_4 + H_2O$$

Variscite (insoluble compound)

The greatest phosphate availability is at about pH 5.8 - 6.5.

Calcium deficiency

Calcium deficiency occurs only on very acid soil (< pH 5.0) or where excessive quantities of potassium or magnesium have been used. Poor plant growth on these soils is usually due to excess soluble aluminum, manganese and/or iron rather than inadequate calcium. Excess levels of aluminium in acid soils cause poorly-developed root systems, with many adventitious roots near the soil surface.

Boron deficiency

Its deficiency is commonly observed in light-textured acidic soils, in soils with high amount of calcium carbonate ($CaCO_3$) or oxides and hydrous oxides of iron (Fe) and aluminium (Al), and in soils with low organic-matter content. Availability of boron is generally low in acid soils, due to its fixation through adsorption by aluminum (Al) and iron (Fe) oxide minerals present in acid soil and leaching from light-textured acidic soil.

Mo deficiencies

Molybdenum is also less available at low pH. Deficiency of Mo is quite common in acid sandy soils and acid peats. In contrast, in acidic soils (pH<5) molybdenum availability decreases as anion adsorption to soil oxides increase. Generally, in acidic soil MoO_4^- may adsorb onto positively charged metal oxides (Fe, Al, Mn), clay minerals, dissolved organic compounds and carbonates. The adsorption of molybdenum onto positively charged metal oxides is strongly pH dependent with maximum adsorption occurring at low pH.

3. Microbial activity

Acid soil conditions impact negatively on soil biological activity. Evidence of this is the retardation in the breakdown of surface-applied organic matter on acid soils. Generally, bacteria and actinomycetes are highly active at moderate to high pH values, but their activities are hampered when the soil pH drops below 5.5. Fungi may grow well in a wide range of soil pH, but predominating in acidic condition.

The microbial activities, affect the bioavailability of both macronutrients and micronutrients, as these organisms are responsible for conversion of unavailable forms of nutrients to their available forms. Most soil microbes thrive in a range of slightly acidic pH (6–7) due to the high bioavailability of most nutrients in that pH range. The population of bacteria that decomposes organic matter declines and their activity is hindered in highly acidic soil, which results in the accumulation of organic matter and the bound nutrients, particularly nitrogen.

Low pH in top soils may affect microbial activity, most notably decreasing legume nodulation. This is brought about largely by the acidity making aluminium available at levels toxic to Rhizobium. Acidity will however affect all microbial cells and the biochemical processes which they mediate. Low soil pH slows the biological transformation of ammonium to nitrate.

Management of acid soils

The first step in the management of acid soils is to identify the extent and severity of the problem. With careful sampling of fields, soil tests can determine the extent and severity of soil, acidity, the rate of lime required, and provide an estimate of crop response to lime. The fertility status of acid soils is very poor and under strongly to moderately acidic soils the plant growth and development get affected to a great extent. The amendments have the ability to reduce the H^+ ions in the soil solution. The amendments were able to bring an increase in soil pH and decrease in exchangeable acidity by neutralizing the acidity.

Liming

One of the most important thing in managing acid soils is liming. Liming is the best measure to correct soil acidity. Liming is a process in which lime, such as calcic limestone ($CaCO_3$); dolomite limestone ($CaCO_3.MgCO_3$); quicklime (CaO), hydrated or slaked lime $Ca(OH)_2$, or silicate salts are added to a soil to neutralize soil acidity, which raise soil pH to a certain level. The main function of liming materials is to produce OH^- ions which neutralize soil acidity (exch. Al^{3+} and H^+).

$$CaCO_3 + H_2O \rightleftharpoons Ca^{2+} + HCO_3^- + OH^-$$

$$CaSiO_4 + 2H_2O \longleftrightarrow Ca^{2+} + H_2SiO_3 + 2OH^-$$

Lime decreases the acidity level of soil (increases pH) by changing the hydrogen ions (H^+) of soil into water and carbon dioxide (CO_2) molecule. One calcium ion (Ca^{++}) from the lime replaces two hydrogen ions (H^+) of soil complex. In addition, one carbonate ion (CO_3^{2-}) reacts with water molecule (H_2O) to form bicarbonate ion (HCO_3^-). These react with hydrogen ion (H^+) to form H_2O and CO_2. Thus, the pH of soil increases due to the concentration of hydrogen ions (H^+) has been reduced, shown in fig 6.2.

Fig. 6.2: Reduction of soil acidity (or H^+ ions) by lime

However, liming is not a remedy for **subsoil acidity** and it is not always economically feasible. Therefore, the most appropriate strategy to overcome from subsoil acidity is the addition of gypsum.

Reaction of liming material in soil

When lime is added to an acid soil, a number of events occur, most of them occurring simultaneously. The chemistry to liming is quite simple. Hydrogen ions (H^+) are attracted to soil and organic material which have a negative charge. When lime is applied, these hydrogen ions are exchanged for calcium or magnesium (Ca^{2+} or Mg^{2+}) ions which have a greater positive charge. This helps to neutralize the acidity of the soil. The free hydrogen ions are taken out of the solution. This also helps to increase the pH.

The following reactions will occur:

(1) Lime is dissolved (slowly) by moisture in the soil to produce Ca^{2+} and hydroxide (OH^-):

$$CaCO_3 + H_2O \text{ (in soil)} \longrightarrow Ca^{2+} + 2OH^- + CO_2 \text{ (gas)}$$

(2) Newly produced Ca^{2+} will exchange with Al^{3+} and H^+ on the surface of acid soils:

$2Ca^{2+}$ + [Soil Particle — Al^{3+}, H^+] ⟹ [Soil Particle — Ca^{2+}, Ca2+] + Al^{3+} + H^+

(3) Lime-produced OH^- will react with Al^{3+} to form solid $Al(OH)_3$, or it will react with H^+ to form H_2O:

$$3OH^- + Al^{3+} \longrightarrow Al(OH)_3 \text{ (solid)}$$

$$OH^- + H^+ \longrightarrow H_2O$$

Thus, above reactions clearly shows that, liming eliminates toxic Al^{3+} and H^+ through the combination with OH^- ion. Excess OH^- from lime will raise the soil pH, which is the most recognizable effect of liming. Another benefit of liming is the added supply of Ca^{2+}, as well as Mg^{2+} if dolomite [$Ca,Mg(CO_3)_2$] is used.

Liming materials

A liming material is a substance that contains calcium (Ca) or magnesium (Mg) or both (Ca and Mg) and that neutralizes excess hydrogen ions and raises the soil pH that will neutralize soil acidity. Common liming materials are the oxides, hydroxides, carbonates, and silicates of calcium (Ca) or magnesium (Mg) or Ca-Mg mixtures. Commercial limestone and dolomite limestone are the most available and widely used liming material. Carbonates, oxides and hydroxides of calcium and magnesium are referred to as agricultural lime. The bulk of agricultural lime comes from ground limestone, which can be calcite ($CaCO_3$) or dolomite ($CaCO_3$, $MgCO_3$).

Liming materials are effective when they:

- Have capacity to remove H^+ and Al^{3+} off of exchange sites (potential acidity)
- Have ability to neutralize H+ in solution (active acidity)
- Have a desirable proportion of cations, mostly calcium or magnesium.
- Are economical.

Kinds of liming materials

The carbonates, oxides, hydroxides and silicates of Ca and or Mg are usually used as liming materials for the correction of soil acidity.

(i) **Ground limestone.** Ground limestone usually contains $CaCO_3$ (calcite) and some impurities. It is the most suitable calcium (Ca) containing liming material where soil pH is low (<6.0) and a large quantity of lime is required to neutralize the acid and increase soil pH to the desired level.

(ii) **Dolomite** A common grounded dolomite containing both Ca and Mg (40 % $MgCO_3$ & 60% $CaCO_3$), applications are appropriate where soil Mg levels are low to replenish soil Mg reserves. Magnesium limestone serves as a source of Mg, and is the most effective way to improve soil Mg levels where lime is required.

(iii) **Calcium oxide** (CaO) A caustic white powder created when $CaCO_3$ is burnt, also called quicklime or burned lime. It has been exposed to high temperature to remove carbon dioxide. Pure calcium oxide has a Calcium Carbonate Equivalent (CCE) of 179% and reacts quickly. Burned lime must be handled carefully as it quickly reacts with water creating hydrated lime and releasing large amounts of heat.

(iv) **Calcium hydroxide** [$Ca(OH)_2$], also called hydrated lime or slaked lime. This is a very fast acting and powdery lime material. It is an effective liming material, but difficult to handle. Care should be taken when using calcium hydroxide, as is caustic and can burn plants when applying to soils where crops are already established. Finely ground hydrated lime can have a CCE or Neutralizing Value (NV) of 120-135% and can quickly rise soil pH.

The other liming sources are marl, oyster shells and several industrial by-products like steel mill slag, blast furnace slag, lime sludge from paper mills, pressmud from sugar mills, etc equally effective as ground limestone and are also cheaper.

(a) **Marl**. – Marl is an unconsolidated material composed of sea shell fragments, and calcium carbonate with varying amounts of silt, clay and organic matter as impurities. They are often found around coastal areas. Marls tend to react similarly to ground limestones.

(b) **Basic Slag** ($CaSiO_3$) – It is a byproduct of the steel industry. Can also contain Mg and P depending on the source of Fe ore used to make steel. Basic Slag having neutralizing value ranges from 60 - 80%.

(c) **Fly Ash** ($CaO + SiO_2$) – it is a byproduct of coal combustion with highly variable calcium carbonate equivalent. Fly ash, is a low- density amorphous ferro-alumino silicate containing by product, which also improves pH and nutrient availability.

(d) **Basic slag** – Basic slag is superior to calcium oxide or carbonates for amending the acid soils because they carry calcium silicate.

Why Gypsum is not considered as a liming material, though it contains calcium?

Gypsum is not considered as liming material because when it applied to an acid soil, it gets dissociated into calcium (Ca^{2+}) and sulphate (SO_4^{2-}) ions.

$$CaSO_4 \rightleftharpoons Ca^{2+} + SO_4^{2-}$$

This sulphate (SO_4^{2-}) ions reacts with soil moisture and produces sulphuric acid, which tends to increases acidity instead of reducing soil acidity.

$$SO_4^{2-} + H_2O \longrightarrow H_2SO_4$$

The Ca^{2+} ions in the gypsum after dissociation will result in the replacement of adsorbed aluminium (Al^{3+}) ions and reduces the level of aluminium in both soil solution and exchange complex, which results in further lowering of soil pH. Therefore, gypsum is not considered as a liming material.

Efficiency of Liming Materials

Lime score is a quality index used to express the efficiency or effectiveness of liming materials for neutralizing soil acidity and is based on the following factors because of their varying neutralizing capacity of various liming materials.

(i) Calcium Carbonate Equivalent (CCE) or Neutralizing Value (NV) of liming materials

(ii) Purity of liming materials and

(iii) Fineness of liming materials.

(i) Calcium Carbonate Equivalent (CCE) or Neutralizing Value (NV) of liming materials

The calcium carbonate equivalent (CCE) is an important measurement in determining how much lime should be applied to the soil. The calcium carbonate equivalent (CCE) or neutralizing value is defined as the capacity of the liming material to neutralize acidity of an acid soil, when it is fully dissolved. Which is expressed as a weight (percentage) of pure calcium carbonate is taken as the standard against which liming materials whose calcium carbonate equivalent (CCE) is measured. The calcium carbonate equivalent (CCE) or neutralizing value (NV) is expressed as a percentage of the neutralizing value of pure $CaCO_3$, which gives a value of 100 %, because pure $CaCO_3$ has molecular weight of 100.

$$\text{CCE of a liming material (\%)} = \frac{\text{Molecular weight of } CaCO_3}{\text{Molecular weight of liming materials whose CCE is to be measured}} \times 100$$

As CCE value increases, the acid neutralizing power of the lime increases. The lower the CCE value, the more lime will need to neutralize the soils acidity.

Commonly used liming materials and their relative neutralizing values are given in Table 6.2.

Table 6.2: Calcium Carbonate Equivalent or Neutralizing Value of some liming materials

Liming materials	Chemical name/ Chemical composition	CCE or Relative neutralizing value (%)
Burnt lime	Calcium oxide (CaO)	179
Slaked/ Hydrated lime	Calcium hydroxide [$Ca(OH)_2$]	136
Calcite/ Limestone	Calcium carbonate ($CaCO_3$)	100
Dolomite	Calcium magnesium carbonate [$CaMg(CO_3)_2$]	109
Basic slag	Calcium silicate ($CaSiO_3$)	86

(ii) Purity of liming materials. The more pure of the liming material, have higher will be its efficiency to reclamation the acidity of a soil.

(iii) Fineness of liming materials. The degree of fineness indicates the speed with which lime materials will neutralize soil acidity. Lime fineness is measured by the proportion of processed agricultural lime which passes through a sieve with a different mesh sizes. The standard mesh size numbers indicate the number of wires per inch. Thus, higher mesh size numbers signify smaller holes, which limit passage to finer particles. A 60-mesh sieve, which is the standard for comparisons of lime fineness and efficiency rating of 100%, is assigned. Liming materials with a smaller particle size will neutralize soil acidity more quickly than materials with similar CCE, but larger particle size. Because lime dissolves very slowly, it must be finely grind to increase its reactive surface area for contact with the soil and to effectively neutralize soil acidity.

Lime requirement and Liming factor

Lime requirement for an acid soil may be defined as the amount of $CaCO_3$ or any other liming material, required to raise the pH of the soil upto a desired value under field conditions to attained optimum pH for proper crop growth, is known as lime requirement. This value is usually in the range of pH 6.0 to 7.0, since this is an easily attainable value within the optimum range of most crop plants. It depends mainly on soil pH, its buffering capacity, chemical composition of liming material and fineness of liming material. Highly buffered soils such as clay, peat and muck soils will have less amount of lime requirement than that of coarse textured soils with low clay and organic matter.

For determination of lime requirement, the buffer method of *Shoemaker, McLean and Pratt* (SMP, 1961) is widely used in India. The SMP buffer method is based on a generalized relationship between the buffer indicates the lime requirement and pure $CaCO_3$ incubation, which can be observed from the Table 6.3.

Liming factor

The liming factor may be defined as the factor by which actual amount of lime can be calculated from the estimated theoretical amount of lime. This factor is varies from 1 to 3, but liming factor of 1.5 to 2.0 is generally used. This factor mainly depends on the rate of limestone solution, plant uptake and leaching during the reaction period.

Methods of applying lime under field condition

Lime reacts with the soil only when water is available. Liming materials are however low in water solubility. Therefore, growers should apply lime to a soil before sowing a crop, as early as possible as time and soil moisture can facilitate lime reaction and soil pH adjustment to the target crop.

When the land is ready for line sowing the furrow are opened at the recommended distance for a particular crop. Powdered lime @ 3-5 t ha^{-1} for desired area of land is applied in an open furrow. It is mixed with soil by feet and then recommended dose of fertilizer is mixed with soil. Seeds are then sown and covered with soil. Lime recommendations are customarily made on the assumption that the liming material will be incorporated to the tillage depth represented by the sample submitted to the laboratory, which is most often 15 cm (6 inches). Liming of acid soils is usually required in every three to five years, depending on several factors like management, rainfall, soil characteristics, etc.

Table 6.3: Lime requirement (tons of pure $CaCO_3$ per acre) to bring the soil to indicated pH.

pH of soil buffer suspension (Field Soil Sample)	Lime required to bring the soil to indicated pH (tones per acre)		
	pH 6.0	pH 6.4	pH 6.8
6.7	1.0	1.2	1.4
6.6	1.4	1.7	1.9
6.5	1.8	2.2	2.5
6.4	2.3	2.7	3.1
6.3	2.7	3.2	3.7
6.2	3.1	3.7	4.2
6.1	3.5	4.2	4.8
6.0	3.9	4.7	5.4

pH of soil buffer suspension (Field Soil Sample)	Lime required to bring the soil to indicated pH (tones per acre)		
	pH 6.0	pH 6.4	pH 6.8
5.9	4.4	5.2	6.0
5.8	4.8	5.7	6.5
5.7	5.2	6.2	7.1
5.6	5.6	6.7	7.7
5.5	6.0	7.2	8.3
5.4	6.5	7.7	8.9
5.3	6.9	8.2	9.4
5.2	7.4	8.4	10.0
5.1	7.8	9.1	10.6
5.0	8.2	9.6	11.2
4.9	8.6	10.1	11.8
4.8	9.1	10.6	12.4

Effect of over liming

In temperate region soils there tends to be in a little danger from adding too much lime, but this is not true of tropical soils. In fact, liming of most tropical soils is better viewed as calcium fertilization than pH adjustment, and the target pH should probably not exceed from desired level. When large amounts of lime is applied it leads to one or many of these following causes :

(i) In addition, soils become deficient in nutrients such as iron, manganese, copper and zinc due to decreasing their solubility.

(ii) The phosphorous and potassium availability will also be reduced.

(iii) However, over liming acid soils often has resulted in temporary boron deficiency, especially when liming to pH levels above 7.0.

(iv) Due to application of lime in excess, the incidence of diseases like scab in root crops will be increased. [Note: Club root disease of cole crops can be reduced with the application of lime].

(v) Over liming can cause a destabilization of the soil structure which, in turn, causes soil aggregates to break apart resulting in reduced permeability and lack of adequate drainage.

(vi) When large amount of lime are applied to sandy textured and low in humus soil (possessing a poor buffering capacity), injury to plant growth may occur.

Correction of over liming

Application of manures, green manuring, composts, phosphate fertilizers and a mixture of micronutrients may reduce lime injury / over liming.

Beneficial effects of liming

Liming is the most important and most effective practice to ameliorate soil acidity constraints for optimal crop production. The practice of well-planned and execution of liming under these situations is fundamental to improve soil fertility and for increasing crop yields on acid soils. This in turn helps to reduce crop production risks associated with soil acidity, as liming promotes nutrients use efficiency, especially of phosphorus.

The main effects of liming on acid soil are:

(i) It maintains the deficiency of Ca and Mg, by adding both essential nutrients.

(ii) Liming reduces the toxicity of aluminium, manganese and iron.

(iii) Lime increases the availability of nitrogen by enhancing the decomposition of organic matter through greater microbial activity at higher pH.

(iv) Liming is a common practice for reducing the mobility of heavy metals because it does not only support the plant nutrition by calcium, but it improves the soil properties and is often used to decrease the heavy metals mobility and hence reduces toxicity in soil.

(v) Improved symbiotic N fixation by legumes.

(vi) Liming increases Ca^{2+} concentrations and ionic strength in the soil solution, causing clay flocculation and thus an improvement in soil structure and hydraulic conductivity.

(vii) Calcium released from applied lime in soil has been reported to enhance plant resistance to several plant pathogens.

References

Havlin, J. L., Beaton, J. D., Tisdale, S. L. and Nelson, W. L. 1999. Soil Fertility and Fertilizers 6th Ed. Prentice Hall. Upper Saddle River, New Jersey.

Maji, A. K., Redyy, O. G. P. and Sarkar, D. 2012. Acid Soils of India– Their Extent and spatial variability. NBSS Publication No. 145, NBSS&LUP, Nagpur, pp. 1-138.

Mandal, S. C., Sinha, M. K. and Sinha, H. 1975. Acid Soils of India and Liming, Technical Bulletin No. 75. ICAR, New Delhi.

Sharma, P. D. and Sarkar, A. K. 2005. Managing Acid Soils for Enhancing Productivity. Technical Bulletin No. 23, NRM Division. Krishi Anusandhan Bhawan II, Pusa Campus, New Delhi.

Shoemaker, H. E., McLean, E. O. and Pratt, P. F. 1961. Buffer methods of determining lime requirement of soils with appreciable amounts of extractable aluminium. Soil Science Society of America Proceedings, 25: 274-277.

7

Eroded and Compacted Soils and Their Reclamation and Management

Reshma Shinde, Shikha Verma and P. K. Sarkar

Introduction

Soil is one of the crucial natural resources that support life on the earth and controls the economic conditions of the nation. Soil erosion has become a serious problem in both rainfed and irrigated areas of India thus it of vital importance to raise awareness about soil erosion and their reclamation / management methods, so that future land management decisions can lead to more sustainable and resilient agricultural systems. In India almost 130 million hectares of land, i.e., 45 % of total geographical surface area, is affected by serious soil erosion through gorge and gully, shifting cultivation, cultivated wastelands, sandy areas, deserts, and water logging.

In India, the soil is eroded at an average annual rate of 16.35 tonnes per hectare which means 5334 million tonnes per year for the country as a whole. Out of this, about 29 per cent is permanently lost to the sea, while 61% is simply transferred from one place to another, and the remaining 10% is deposited in reservoirs (which mean the storage capacity is reduced by 1-2 per cent annually (Narayana and Ram Babu, 1983)). Singh *et al.* (1990) estimated that the average annual soil loss is about 15.2 tonnes per hectare and at national level it amounts to about 4978 million tonnes annually. The annual erosion rates vary from region to region. In dense forests covers, snow-clad cold deserts and arid regions of western Rajasthan, the annual erosion rates are less than 5 tonnes per hectare. On the other hand, about 64 per cent of the total soil is contributed by highly to very severely eroded areas, such as the Shiwalik hills (annual rate is more than 80 tonnes per hectare), the Western Ghats, black and red soil regions, ravines and other gully eroded areas and the northeastern region. For the first time, Gurmel Singh prepared map of soil erosion in India (Fig. 7.1), on the basis of iso-erosion lines i.e. line joining the place of same erosion rates. The levels of soil erosion are classified by the degree of severity in Table 7.1. It shows that moderate erosion of 5–10 tons per ha (per year) is the largest category affecting 43 % of the total area affected by soil erosion. About 1.4 billion tons of soils are lost annually due to moderate erosion, and 1.6 billion tons due to high erosion. The total annual soil losses are estimated at about 5 billion tons. Using the annual soil specific erosion rates provided by the ICAR-CSWCRTI the estimated loss of major nutrients is nearly 74 million tons annually due to erosion in India. On an average, every year, the country loses 0.8 million tons of nitrogen, 1.8 million tons of phosphorus, and 26.3 million tons of potassium (State of the Environment, 2001). The rivers Ganga and Brahmaputra carry the maximum sediment load annually, about 586 and 470 million tons, respectively. Between 6000 and 12,000 million tons of fertile soil are eroded annually and much of it is deposited in the reservoirs leading to a reduction in their storage capacity by 1–2 % (State of the Environment, 2001).

Table 7.1: Levels of Severity of Soil Erosion in India (Source: Singh, G., 1990)

Severity of Erosion	Annual Soil Loss Range(tones/hectare)	The Share of Total Affected Area (%)	Annual Loss of Soil (million tons)
Slight	≤5	24	401
Moderate	5-10	43	1406
High	10-20	24	1610
Very High	20-40	5	640
Severe	40-80	3	666
Very Severe	≥80	1	255
Total			4978

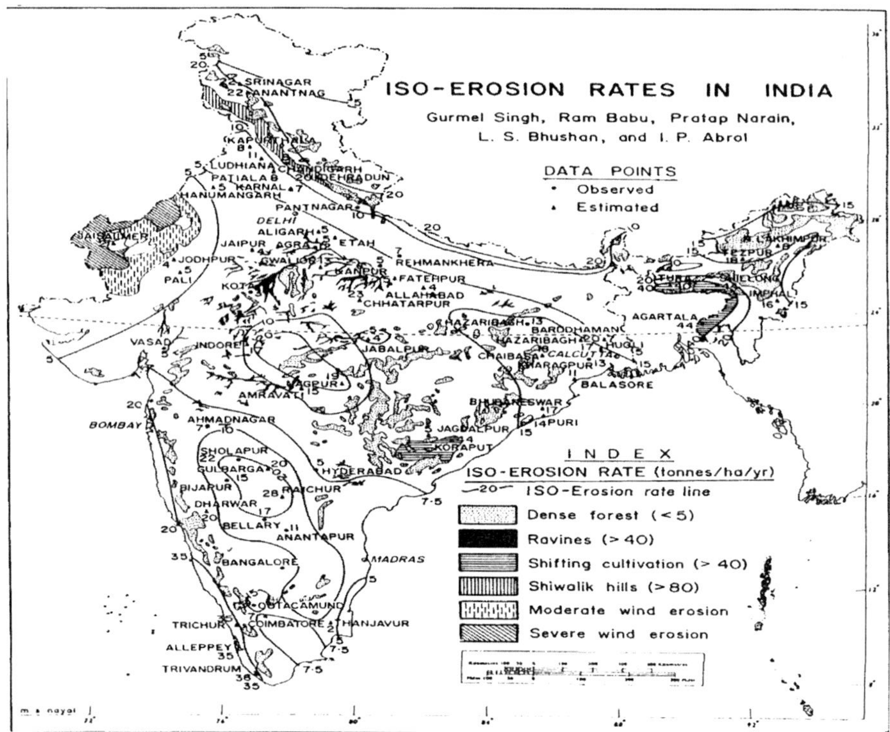

Fig 7.1: ISO erosion Rates in India

Source: http://www.ciesin.columbia.edu/docs/002-413/fig1.gif

Table 7.2: Extent of soil erosion in India, as assessed by different sources

Organizations	Assessment Year	Area (Mha)
National Commission on Agriculture	1976	150
Sehgal and Abrol	1994	162.4
Sehgal and Abrol	1997	167.0
NBSS&LUP	2005	119.19

Eroded soil

Eroded soil means the soil which has been exposed to any kind of erosion which has caused deterioration in its physical and chemical properties. Erosion not only removes upper fertile layer but also causes soil crusting or sealing, soil compaction, poor soil structure, low organic matter, poor drainage and run-off. This all led to formation of eroded soil. The erodation of soil is caused by both natural and anthropogenic activities. Now a day's anthropogenic activities play a major role for formation of eroded soil.

The main causes for formation of eroded soil

- Crop Residue Burning and Inadequate Organic Matter Inputs,
- Overgrazing, Deforestation and Careless Forest Management
- Faulty irrigation methods
- Poor Irrigation and Water Management
- Poor Crop Rotations
- Low and Imbalanced Fertilization
- Excessive Tillage and Use of Heavy Machinery
- Pesticide Overuse and Soil Pollution
- Population increase
- Urbanization, Industrialization and Mining
- Land Fragmentation, Land Shortage and Poor Economy
- Natural processes e.g. earthquakes, tsunamis, droughts, avalanches, landslides, volcanic eruptions, floods, tornadoes, and wildfires

Properties of eroded soils

- Eroded soils are usually shallower and plant roots have less volume to exploit for water and nutrients compared with less eroded soils.
- The permanent reductions in crop productivity of eroded soils are caused primarily by changes in soil physical properties.
- In eroded soils, there is decrease in water retention because of the removal of clay and silt size particles that occurs with erosion
- In eroded conditions, soil water is held strongly, and additional energy is required by plants for nutrient uptake when compared with uneroded conditions.
- In eroded soils not only upper fertile layer is removed but also soil crusting or sealing and soil compaction is observed.
- The erosion of soils causes destruction in soil structure and texture.
- These soils are deprived of organic matter and have low fertility status.
- These soils have poor drainage due to which run-off is more.

- The eroded soil are usually having low permeability and infiltration rate due to this there is decrease in the available soil water capacity of eroded soils.
- Eroded soils consist of the rill and gully formation

The soil characteristics which determines the vulnerability of soil to erode is as following:

i) **Soil Texture:** Soils that contain high proportions of silt and very fine sand are the most erodible and are easily detached and carried away. The erosion potential of soil decreases as the percentage of clay or organic matter increases; clay acts as a binder and tends to limit erosion potential. Most soils with high clay content are relatively resistant to detachment by rainfall and runoff. Once eroded, however, clays are easily suspended and settle out very slowly. Sand, sandy loam and loam-textured soils tend to be less erodible than silt, very fine sand, and certain clay-textured soils.

ii) **Organic Matter Content:** Soils with low organic matter content are more erodible than soils with high organic matter. Organic matter creates a favorable soil structure, improving its stability and permeability. This increases infiltration capacity, delays the start of erosion, and reduces the amount of runoff. The addition of organic matter increases infiltration rates (and, therefore, reduces surface flows and erosion potential), water retention, pollution control, and pore space for oxygen.

iii) **Soil Structure:** Organic matter, particle size, and gradation affect soil structure, which is the arrangement, orientation, and organization of particles. Generally, soils with faster infiltration rates, higher levels of organic matter and improved soil structure have a greater resistance to erosion. Structure less soils, sandy soils, soils low in organic matter and water holding capacity, are more vulnerable to erosion.

iv) **Soil Permeability:** Soil permeability refers to the ease with which water passes through a given soil. Well-drained and well-graded gravel and gravel mixtures with little or no silt are the least erodible soils. Their high permeability and infiltration capacity helps prevent or delay runoff.

Soil erosion

Soil erosion in simple way is the process of detachment, transportation and deposition of soil particles from land surface. The agencies or the energy sources involved in the process of soil erosion are mainly water, wind, gravity, glaciers, sea waves, human beings and animals (Judson, 1965; Merritt et al., 2003). Soil erosion is defined as the detachment and transportation of soil particles from one place to another and deposition elsewhere through the action of the wind, water, coastal waves, glacier, gravity and other forces. Soil erosion begins with detachment, which is caused by breakdown of aggregates by raindrop impact, sheering or drag force of water and wind. Detached particles are transported by flowing water or wind and deposited when the velocity of water or wind decreases by the effect of slope or ground cover. Thus Soil erosion involves three steps, viz i) detachment of soil particles from the main soil bay ii) their transportation by splashing, floating, rolling, dragging and iii) deposition at another place.

Table 7.3: The state wise extent of soil erosion from wind as well water

Sr.no	Name of the States	Water Erosion (1000 ha)	Wind Erosion (1000 ha)
1	Andhra Pradesh + Telengana	11518	0
2	Arunachal Pradesh	2372	0
3	Assam	688	0
4	Bihar+ Jharkhand	3024	0
5	Goa	60	0
6	Gujarat	5207	443
7	Haryana	315	536
8	Himachal Pradesh	2718	0
9	Jammu &Kashmir	5460	1360
10	Karnataka	5810	0
11	Kerala	76	0
12	Madhya Pradesh+ Chhattisgarh	17883	0
13	Maharashtra	11179	0
14	Manipur	133	0
15	Meghalaya	137	0
16	Mizoram	137	0
17	Nagaland	390	0
18	Orissa	5028	0
19	Punjab	372	282
20	Rajasthan	3137	6650
21	Sikkim	158	0
22	Tamil Nadu	4926	0
23	Tripura	121	0
24	Uttar Pradesh+ Uttarakhand	11392	212
25	West Bengal	1197	0
26	Delhi	55	0
27	Union Territories	187	0
	Total	93680	9483
	Total(Million ha)	93.68	9.48

(*Source*: NBSS&LUP, 2005)

Soil erosion accounts for 87% of the total degraded land in India and is one of major factor causing land degradation. Soil erosion by rain and transportation of soil particles through rivulets that takes place in hilly areas causes severe landslides and floods. Thus Soil erosion in India is amongst the leading areas of concern as it affects cultivation and farming in the country in adverse and unfavorable ways.

Soil erosion is broadly classified in two categories

i) natural or geologic or the normal soil erosion;

ii) accelerated or abnormal soil erosion.

1. **Natural or geologic soil erosion:** The natural / geologic soil erosion takes place, as result of the action of wind, water, sea waves, glacier and gravity and it takes place at such slow rates that loss of soil is compensated for the formation of new soil under natural weathering process. Geological erosion may be considered as part of natural soil forming process which results in the existing form and distribution of soils. This kind of erosion does not pose any problem.

2. **Accelerated or the abnormal soil erosion:** Accelerated soil erosion is caused by the disturbances of humans beings (indiscriminate cutting of forest, removal of vegetative covers, overgrazing , faulty agriculture practices, constructing roads and buildings and etc.). When the soil erosion exceeds the normal rate and becomes unusually destructive and unproductive, it is called as Accelerated soil erosion. In this erosion the removal of soil takes place at a faster rate than that of soil formation.

   Factors affecting soil erosion: Numerous factors affect soil erosion depending upon the erosive forces of wind and water, the local conditions with regard to the physical, chemical and biological nature of the soil and etc. The major factors influencing the soil erosion are:

i) climate ii) vegetation iii) topography iv) soil and v) biotic activity.

Hence soil erosion can be expressed as a function of:

$$E = f\,(C, T, S, V, B\ldots) \text{ ------------eq1}$$

Where, E represents the rate of erosion and C, T, S, V and B Stands for climate, topography, soil, vegetation and biotic activity respectively. The vegetation and to some extent soil factors can be controlled. Climatic factors and topographic factors, except slope length are beyond the control of human beings.

i) **Climate:** The major climatic factor influencing runoff and soil erosion are rainfall, temperature, and wind, of which rainfall is the most important. The soil erosion depends on the intensity, kinetic energy, amount, duration and frequency of rainfall. The intensity of rainfall is important since a heavy rainfall with low intensity may cause less erosion as compared to a low rainfall with high intensity. The wind also influences the angle and impact of raindrops.

ii) **Vegetation:** The nature of vegetative cover plays an important role in controlling erosion. The thick canopy of vegetation acts as shields on the soil surface and protects it from the impact of falling raindrops. The vegetation root system of helps to bind the soil particles together, improves infiltration capacity of soil and reduces the velocity of runoff.

iii) **Topography:** The land slope, its degree and length, is important in determining the extent of soil erosion. Higher the degree of slope and more its length, more is the soil erosion due to increase in velocity of water running down the slope. If the velocity of water flow is doubled, its erosive power increases four times and carrying capacity 64 times. Similarly the velocity of wind and storms also affects soil erosion in the arid regions. Large sand dunes are transported from one place to another within no time due to wind action.

iv) **Soil:** Erodibility of soil is influenced by its physical and chemical properties. It is influenced by soil texture, structure, organic matter, nature of clay and amount and kind of salts present. Erosion is more in fine textured soils. The organic matter helps in binding soil particles and improves water holding capacity of the soil. The three most significant soil charcterstics which influences the soil erosion are i) infiltration capacity, ii) structural stability, and iii) antecedent soil moisture.

v) **Biotic Interference and mass landslides:** Soil erosion due to loosening and cutting of soil on the river bank is accelerated during floods. An earthquake shakes and de-stabilizes the soil surface, thereby rendering it prone to erosion in the form of mass landslides. Faulty agricultural practices without proper care of resources conservation like shifting cultivation and free range grazing by cattle's causes' disturbance and removal of top fertile soil. This result in soil fertility depletion and makes land un suitable for cultivation. Construction of roads and railway tracks in hills produce mass wasting.

Water Erosion

The soil erosion caused by water as an agent is called water erosion. In water erosion, the water acts as an agent to dislodge and transport the eroded soil particle from one location to another. Water erosion (Fig. 7.2).occurs when excess rainfall generates runoff to transport the soil away from its locations. Soil erosion by water results largely in the loss of top soil and terrain deformation.

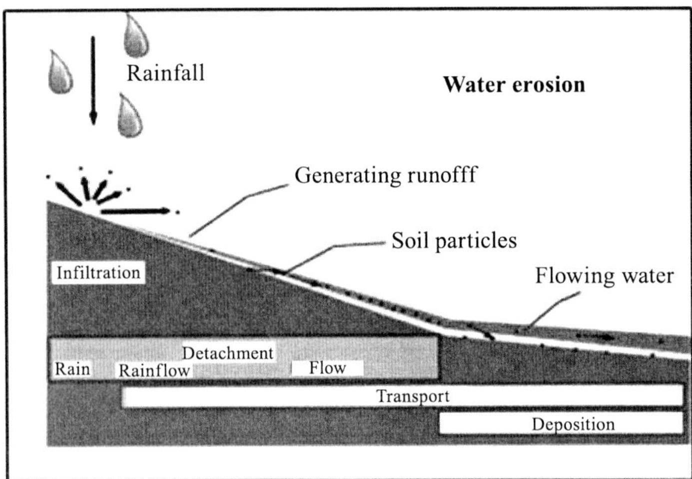

Fig. 7.2: Process of water erosion

Extent of Water Erosion: The extent of water erosion in different states of Indian by National Bureau of Soil Survey & Land Use Planning (NBSS&LUP 2005) is presented in Table 7.3. In India, 93.48 Mha areas are affected by water erosion which accounts for 28% of the total area of India. However, the water erosion extent estimated by different sources varies from 87 to 111 Mha in India.

Causes of Water Erosion: Water erosion is due to the dispersive action, and transporting power of water. Water erosion of soil starts when raindrops strike bare soil peds and clods, resulting the finer particles to move with the flowing water as suspended sediments. The soil along with water moves downhill, scouring channels along the way. Each subsequent rain erodes further amounts of soil until erosion has transformed the area into barren soil. Water erosion may occur due to the removal of protective plant covers by tillage operation, burning crop residues, overgrazing, overcutting forests etc. inducing loss of soil.

Types of water erosion

1. **Splash Erosion:** The detachment and splash or transport of the soil particles occurring as result of impact of falling raindrops is called Splash or raindrop erosion. The kinetic energy of falling raindrop dislodges the soil particle and the resultant runoff transports soil particles. Splash erosion is the first stage of soil erosion by water. It occurs when raindrops hit bare soil. The explosive impact breaks up soil aggregates so that individual soil particles are 'splashed' onto the soil surface. The splashed particles can rise as high 0.60 meter above the ground and

move up to 1.5 meter from the point of impact and the effect of this type of erosion are solely local. The particles block the spaces between soil aggregates, so that the soil forms a crust that reduces infiltration and increases runoff.

2. **Sheet Erosion:** Sheet erosion is the removal of more or less uniform thin layers or sheet of soil by running water from sloping land. This action in called skimming and is prevalent in the agricultural land. It results in loss of the finest soil particles that contain most of the available nutrients and organic matter in the soil. This type of soil erosion is mainly responsible for loss of soil productivities. Soils most vulnerable to sheet erosion are overgrazed and cultivated soils where there is little vegetation to protect and hold the soil. Early signs of sheet erosion include bare areas, water puddling as soon as rain falls, visible grass roots, exposed tree roots, and exposed subsoil or stony soils. Soil deposits on the high side of obstructions such as fences may indicate active sheet erosion. Vegetation cover is important to prevent sheet erosion as it protects the soil, interferes water flow and encourages water to infiltrate into the soil.
3. **Rill Erosion:** It is the advance stage of sheet erosion which occurs when surface water concentrates in depressions or low points through paddocks and erodes the soil. Rills formation is the intermittent process of transforming to gully erosion and thus it is is initial stage of gully formation. The rills are shallow drainage lines less than 30cm deep and 50 cm wide. Rill erosion is common in bare agricultural land, particularly overgrazed land, and in freshly tilled soil where the soil structure has been loosened. The rills can usually be removed with farm machinery. Rill erosion is mostly occurs in alluvial soil and is quite frequent in Chambal river valley in India.
4. **Gully Erosion:** The advance stage of rill erosion and it occurs when rills continue to extend in width, depth and length. Gully formation are initiated when the depth and width of the rill is more than 50 cm. Gullies are deeper channels that cannot be removed by normal cultivation. Hillsides are more prone to gully erosion when they cleared of vegetation, through deforestation, over-grazing or other means. The eroded soil is easily carried by the flowing water after being dislodged from the ground, normally when rainfall falls during short, intense storms. Depending upon the depth, width and side slope the gullies further divided into 4 classes namely G1, G2, G3 and G4 (Table 7.4).the gullies can be further classified based on shape as U shaped and V shaped gullies. U shaped gullies are usually formed in alluvial valleys where both the surface and subsurface soils are easily eroded. V shaped gullies are formed in those areas where subsurface soils are more resistant than top soils. This is

most common type of gully shape. Gullies reduce the productivity of farmland where they incise into the land, and produce sediment that may clog downstream water bodies.

Table 7.4: Classification of gully erosion

Particulars	Description of symbols of Gully			
	G1	G2	G3	G4
Depth in meter	Upto 1.0	1.0-3.0	3.0-9.0	>9.0
Width in meter	<18.0	<18.0	18.0	>18.0
Side slope (%)	<6.0	<6.0	6.0-12.0	>12.0

5. **Stream Bank Erosion:** Stream bank erosion occurs where streams begin cutting deeper and wider channels as a consequence of increased peak flows or the removal of local protective vegetation. Stream bank erosion is common along rivers, streams and drains where banks have been eroded, sloughed or undercut. Generally, stream bank erosion becomes a problem where development has limited the meandering nature of streams, where streams have been channelized, or where stream bank structures (like bridges, culverts, etc.) are located in places where they can actually cause damage to downstream areas. Stabilizing these areas can help protect watercourses from continued sedimentation, damage to adjacent land uses, control unwanted meander, and improvement of habitat for fish and wildlife.

6. **Land Slides Or Landslip Erosion:** Land slippage also known as mud slide or landslide or mass erosion occurs on wet slopping land as saturated soil along the water flow slips down the hillside or mountains slope. Banks along highways, stream and ocean fronts are often prone to landslides. Landslide is the downward and outward movement of slope forming material composed of natural rocks, soil, artificial fills or combination of these materials. Landslips are smaller masses moving allof sudden while slides are bigger masses moving slowly through initiating as slips major causes of landslides are excavation or undercutting of base of an existing slope, weak geology and lack of vegetative cover on slope of slips. Landslides pose an immense threat to highways, villages and agriculture lands in the Himalayas, peninsular India and elsewhere.

7. **Ravine Formation:** Deep and narrow gullies with abrupt sides are usually called as ravines. Ravines represent a severe erosion hazard resulting from enlargement of rills due to continuous non judicious use of the land. The ravines system consists of almost a parallel set of gullies and is always associated with some river system. In India about 3.7 m ha area is affected by ravines mostly found in the state Uttar Pradesh, Madhya Pradesh, Rajasthan, Gujarat and Bihar.

Wind erosion

Wind erosion is the detachment, transportation and redeposition of soil particles by wind. A sparse or absent vegetative cover, a loose, dry and smooth soil surface, large fields and strong winds all increase the risk of wind erosion. Soil erosion by wind has caused an accumulation of eroded particles in loess, a type of soil which makes up some of the world's most fertile and productive regions. Soil conditions favorable to wind erosion are most commonly found in arid and semi-arid areas where rainfall is low or insufficient and no vegetative cover on the land. The most serious damage caused by wind erosion is the change in soil texture. Since the finer soil particles are subject to movement by wind, wind erosion gradually removes silt, clay and organic matter from the top soil, leaving the coarser soil material.

1. Extent of wind erosion

According to Global Assessment of Human-induced Soil Degradation (GLASOD), 21.6 M ha area of Indian soil is affected by wind erosion, which account for 6% of the total geographical area, while According to ICAR-NBSS&LUP, 9.5 M ha area of Indian soil is affected by wind erosion (Table 7.3.). It is mainly found arid and semi-arid areas of India

2. Process of wind erosion

Wind erodes the soil in three steps *i.e* the soil particles are carried by the wind in three ways namely saltation, suspension and surface creep.

i) **Saltation:** It is a process of soil movement in a series of bounces or jumps. Soil particles having sizes ranging from 0.05 to 0.5 mm generally move in this process. Saltation movement is caused by the pressure of the wind on the soil particle, and collision of a particle with other particles. The height of the jumps varies with the size and density of the soil particles, the roughness of the soil surface, and the velocity of the wind.

ii) **Suspension:** Suspension represents the floating of small sized particles in the air stream. Movement of such fine particles in suspension is usually started by the impact of particles in saltation. Once these fine particles are picked up by the particles in saltation and enter the turbulent air layers, they can be lifted upward in the air and they are often carried for several miles before being redeposited elsewhere. Dust particles will fall on the surface only when the wind subsides or the rain washes them down.

iii) **Surface Creep**: Surface creep is the rolling or sliding of large soil particles along the ground surface. They are too heavy to be lifted by the wind and are moved primarily by the impact of the particles in saltation rather than by direct force of the wind. The coarse particles tend to move closer to the ground than the fine ones.

Threshold Velocity: Threshold velocity is the minimum wind velocity required to initiate the movement of soil particles. Threshold velocity varies with the soil conditions and nature of ground surface

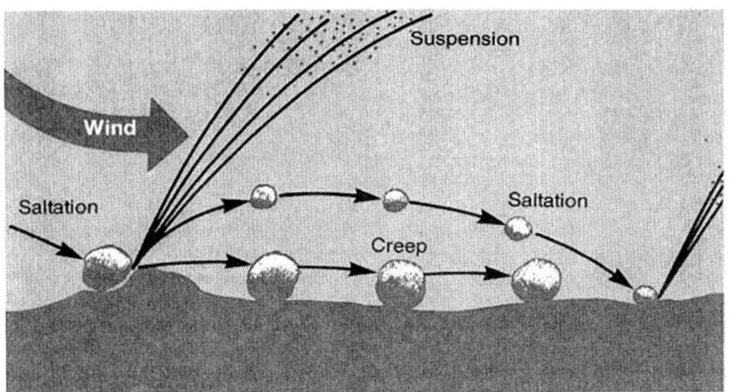

Fig. 7.3: Process of wind erosion
(*Source*:http://www.weru.ksu.edu/new_weru/images/CreepSaltSusp.jpg.)

Effects of soil erosion/ eroded soil

The soil erosion adversely affects the livelihood of the people in one way or other. The major losses and problems occured due to the soil erosion from various agents are listed below.

- Soils eroded by water get deposited on river beds, thus increasing their level and causing floods. These floods sometime have various extreme effects, such as killing human and animals and damaging various buildings.
- Siltation of rivers, irrigation channels and reservoirs, damage to sea coast and formation of sand dunes
- Soil erosion decreases the moisture supply by soil to the plants for their growth. It also affects the activity of soil micro-organisms thus deteriorating the crop yield.
- Top layer of soil contains most of the organic matter and nutrients, loss of this soil reducing soil results in fertility and affecting its structure badly.

- Wind erosion is very selective, carrying the finest particles - particularly organic matter, clay and loam for many kilometers. There the wind erosion causes losses of fertile soils from highly productive farming areas.
- The most spectacular forms are dunes - mounds of more or less sterile sand - which move as the wind takes them, even burying oases and ancient cities.
- Sheets of sand travelling close to the ground (30 to 50 metres) can degrade crops.
- Wind erosion reduces the capacity of the soil to store nutrients and water, thus making the environment drier.
- Thus the main problems caused by soil erosion include risk to food aesthetic security, a decline in esthetic landscape beauty, increase in the probability of flood in flood plains, reduced quality of water, and loss of aquatic biodiversity in rivers and lakes by pollution, eutrophication, decrease in soil fertility and productivity, transformation of land into fallow land not suitable for reforestation, irreversible reduction in arable soil, increase in flooding events, diffuse pollution of river networks and turbidity

Management of eroded soils/ Soil erosion

In Indian condition, the control of soil erosion is a challenging task in the sense that the onset of monsoon often coincides with the *kharif* sowing and transplanting. In this stage of kharif crop when canopy cover is minimal, major part of the land is exposed to the rainfall leading to soil erosion. Eroded Soils can be managed by adopting land management practices and also by changing the pattern of some human activities which accelerate soil erosion. One of best mean is to reduce the erosion of soils and to minimize disturbance in soil. The management of Eroded soil is not only anti erosion and anti runoff approach but it is integrated approach for judicious use of natural resources. The techniques used in controlling soil erosion includes mechanical and agronomical measures, they are as follows:

1. Mechanical measures

a) **Contour Bunds:** Contour bunds, either mechanical or vegetative barrier created across the slope for safe diversion of excess runoff and retaining eroded soil. A study conducted at Doon valleys in the northwestern hills region indicated that contour bunds decreased runoff to the 25%–30% as extent of compared to field bunds [CSWCR&TI Vision,2011].

b) **Graded Bunds:** The graded bunds is a small earthen bund with slight grade constructed across the slope for safe disposal of runoff . it is constructed in region having medium to high rainfall of ~600 mm year^{-1} with 10% slope. In vertisols (of central India), graded broad bed and furrow system of land configuration improves surface drainage and allows better water infiltration. It also facilitates drainage of excess water through grassed waterways. However, the broad bed and furrow system is not as effective for shallower Vertic soils, as it encourages runoff. Runoff and soil loss were lower from broad bed and furrow land surface management practices than from a flat on grade system (Table 7.5). The broad bed and furrow system decreased soil loss to a greater extent (31% to 55%) than its effect on runoff volume (24% to 32%) compared with that of flat on grade system.

Table 7.5: Seasonal rainfall, runoff and soil loss from different land configuration, broad-bed and furrow (BBF) and flat on grade (FOG) (Data source: Mandal et al.2013)

Year	Rainfall (mm)	Runoff (mm)		Soil Loss (ton ha^{-1})	
		BBF	FOG	BBF	FOG
2003	1058.0	163 (15.4%)	214.9 (20.3%)	2.0	2.9
2004	798.2	124 (15.5%)	183.3 (23.0%)	0.7	1.5
2005	946.0	177 (18.7%)	246.0 (26.1%)	1.4	3.1
2006	1513.0	502 (33.2%)	873.0 (57.7%)	3.5	6.4

c) **Bench Terrace and Half Moon Terrace:** The bench terrace are flat beds constructed on hills across the slope and adopted where soil depth is >1.0 m. Half-moon terraces are level circular beds having 1 to 1.5 m diameter cut into half-moon shape on the hill slopes. Beds are used for planting and maintaining saplings of fruit and fodder trees in horticulture/agro-forestry land uses.

d) **Check Dam:** A check dam is a small, sometimes temporary dam constructed across a swale, drainage ditch or waterway to counteract erosion by reducing water flow velocity. A check dam is designed to control the run off velocity of water so that the area below is prevented from eroding. Check dams help not only in reducing the velocity but also the overall amount of water lost as runoff. It thus contributes to not only soil conservation but also to water conservation.

e) **Contour Tillage:** the field is ploughed along the contours and not along the hill slope. It leads to formation of ridges and furrows against the direction of flow and reduces the velocity of water. It also increases the infiltration of water into the soil and more absorption by plants. Thus the total amount of runoff is also reduced. Contour bunding also works in a similar manner.

f) **Terraced Farming:** In Terrace Farming, wide steps are cut around the slopes of hills or physical barriers (trenches, stone bunds, live hedges, or natural strips) are created along the contours to retain rain water and make it infiltrate into the soil and to minimize soil erosion. Fairly flat strips of land developed between barriers where crops are grown. Terraces are more effective than contour plowing alone in minimizing soil erosion and facilitating infiltration of rainfall into the soil. Terrace Farming alters the shape of the slope to produce flat areas that provide a catchment for water.

g) **Water Harvesting Ponds:** water harvesting structures are dug-out embankment type of water harvesting structure used for creating seasonal and perennial ponds by retaining seasonal and perennial runoff at the foot of a micro-watershed for irrigation and fish farming purposes.

h) **Grassed Waterways:** They force storm runoff water to flow down the center of an established grass strip and can carry very large quantities of storm water across a field without erosion. Grassed waterways are also used as filters to remove sediment, but may sometimes lose their effectiveness when too much sediment builds up in the waterways. To prevent this, it is important that crop residues, buffer strips, and other erosion control practices and structures be used along with grass waterways for maximum effectiveness.

i) **Embankments:** As land development happens in hilly countryside, more erosion control on steep slopes and embankments is needed. Especially in areas that experience heavy rainfall. An embankment is a raised structure (as of earth or gravel) used especially to hold back water. Embankments are constructed of a material that usually consists of soil, but may also include aggregate, rock or crushed paving material. embankment prevents the soil from being washed way and hold back water. Embankments constructed across canals make sure that the soil in the surrounding area doesn't get affected.

j) **Retaining Walls:** Retaining walls are structures designed to restrain soil to unnatural slopes. They are used to bound soils between two different elevations often in areas of terrain possessing undesirable slopes or in areas where landscape needs to be shaped severely. Retaining wall is a structure designed & constructed to resist the lateral pressure of soils where there is desired change in ground elevations that exceeds the angle repose of the soil. Retaining walls are commonly used not only for control of erosion, but also to protect shorelines & keep rainwater from seeping into unwanted areas. The basic material used in construction of retaining wall is concrete, which is meant to serve the function of strength & durability. Retaining walls are relatively rigid walls used for supporting the soil mass laterally so that the soil can be retained at different levels on the two sides.

k) **Runoff Diversion Structures**: These are channels that are constructed across slopes that cause water to flow to a desired outlet. They are similar to grassed waterways and are used most often for gully control.

l) **Drop Structures:** Are small dams used to stabilize steep waterways and other channels. They can handle large amounts of runoff water and are effective where falls are less than 2.5 meters

m) **Contour Drain:** It is a temporary ridge or excavated channel, or combination of ridge and channel, constructed to convey water across sloping land on a minimal gradient to periodically break overland flow across disturbed areas in order to limit slope length.

n) **Level Spreader:** A non-erosive outlet to disperse concentrated runoff uniformly across a slope. The level spreader provides a relatively low cost option, which can convert concentrated flow to sheet flow and release it uniformly over a stabilized area.

o) **Pipe Drop Structure / Flume:** A temporary pipe structure or constructed flume placed from the top of a slope to the bottom of a slope. A pipe drop structure or a flume structure is installed to convey surface runoff down the face of unstabilised slopes in order to minimize erosion on the slope face.

p) **Surface Roughening:** Roughening a bare earth surface with horizontal grooves running across a slope or tracking with construction equipment. It helps in the establishment of vegetative cover, reduce runoff velocity, increase infiltration, reduce erosion and assist in sediment trapping.

q) **Chiseling**- This system does not turn the soil over, but rather leaves it rough and cloddy with plenty of crop residue remaining. The soil density and amount of covering depends on the depth, size, shape, spacing, and so on of the chisel blades. The residue and rough, cloddy surface of the soil reduces raindrops impact and reduces runoff velocities thus reducing erosion.

r) **Disking**- This system pulverizes the soil and gives great soil density. The effect is similar to that of chiseling with results also depending on the depth, size, spacing, and so on of the disk blades.

s) **Strip Rotary Tillage:** A strip four to eight inches wide and two to four inches deep is prepared by a rotary tiller, while the rest of the soil is left undisturbed. The soil is conserved because of the crop residues between the tillage strips.

t) **No-till Planting**: This planting system prepares a seedbed 2 inches wide or less, leaving most of the surface undisturbed and still covered with crop residues. The result is a wetter, colder environment that protects the seed and soil with its insulating effect of the surface residue.

u) **Till Planting**: This plowing technique sweeps the crop residues into the area between the rows of crops. Soil density between these rows remains relatively high because of the absence of tillage. This soil is difficult for raindrops to detach and runoff to move.

v) **Annual Ridges**: Also known as permanent ridges or ridge tillage, the annual ridges are formed by using a rolling disk bedder, and planting is done after only minor spring seedbed preparation. The extent of soil conservation depends on the amount of residue left and the row direction. Planting on the contour increases surface residues greatly reduce soil loss.

w) **Geosynthetic Erosion Control Systems (GECS):** The protection of channels and erodible slopes utilizing artificial erosion control material such as geosynthetic matting, geotextiles or erosion matting. There are both Temporary and Permanent Non-Degradable GECS.

Agronomical measures

a) **Strip cropping:** It implies growing of crops in strips parallel to one another. While some strips may be allowed to remain fallow at a time, others might be under a crop. Some of the strips may be used for raising tree crops too. The trees or tall crop strips act as wind breaks or shelter belts. As different crops are harvested at different times, only a limited area is left bare or exposed to soil erosion at a time. Adjoining strips to such areas are at that time under a crop or tree cover thus providing protection to even the bare strip from erosion.

b) **Crop rotation:** Improves the overall efficiency of nitrogen uptake and utilization in the soil. If certain cover crops are planted in the winter, erosion and runoff is prevented when the ground thaws, and nutrients are trapped in the soil and released to the spring crops.

c) **Mulching:** Mulches are loose coverings or sheets of material placed on the surface of cultivated soil. Organic mulches also improve the condition of the soil. As these mulches slowly decompose, they provide organic matter which helps keep the soil loose. This improves root growth, increases the infiltration of water, and also improves the water-holding capacity of the soil. It protects the soil surface from the erosive forces of raindrop impact and overland flow. Mulching assists in soil moisture conservation, reduces runoff and erosion, controls weeds, prevents soil crusting and promotes the establishment of desirable vegetation.

d) **Vegetative Barriers:** The use of vegetative barriers in Himalaya region have decrease runoff by 18%–21% and soil loss by 23%–68% on slopes varying from 2%–8% (Table 7.6). Vegetative barriers of *Guinea* grass, *Khuskhus* and *Bhabar* were effective (after 3–4 years) in reducing soil loss by 6–8 ton ha^{-1}year^{-1} and runoff by 33%–38% (CSWCR&TI Vision, 2011)

Table 7.6: Effect of grass barriers on yield, runoff and soil loss in different slopes of the northwestern hill region

Particulars	Slope (%)					
	2	4			8	
	Guinea Grass	Guinea Grass	Khus khus	Bhabar	Guinea Grass	Khus khus
Runoff (% of total rainfall)	25.8	33.3	35.1	37.9	38.90	40.52
Soil loss (ton ha^{-1}year^{-1})	3.27	6.12	6.72	8.34	9.45	9.87

(*Data Source*: CSWCR&TI Vision 2030)

e) **Shelter Belt and Windbreaks:** Shelter Belt consisting of several rows of trees established at right angle in order to deflect air currents, to reduce the velocity of winds, to protect the leeward areas from desiccating effects of hot winds and wind erosion, and to provide fuel, fodder timber etc. Windbreaks are linear plantings of air-resistant trees and shrubs, which protects fields, homes, canals or other surrounding areas from the strong winds and blowing soil or sand. Wind-breaks should be-raised at right angles to the direction of wind. N - S direction is good compromise.

f) **Agroforestry:** Agroforestry systems are an appropriate management tool for eroded soils, because perennial woody vegetation recycles nutrients, maintains soil organic matter, and protects soil from surface erosion and runoff [Nair, 1993]. Tree vegetation in an agroforestry system serves two major purposes: (i) the fine root system holds soil in place, reducing susceptibility to erosion; and (ii) plant stems decrease the flow velocity of runoff, enhancing sedimentation. Mishra, *et al* (2004) opined that soil erosion can be decreased in alkaline soils with *Prosopis juliflora* and *Casuarina equisetifolia* due to the formation of stable soil aggregates in the surface layers. Kaur, *et al.* (2000) analyzed the role of agroforestry systems *Acacia*, *Eucalyptus* and *Populus* along with rice–berseem (*Trifolium alexandrinum* L.) to improve soil organic matter, microbial activity and N availability and observed that: (i) microbial biomass C and N were greater by 42% and 13%, respectively, in tree-based systems than mono-cropping; (ii) soil organic C increased by 11%–52% due to integration of trees along with crops after 6–7 years.

g) **Conservation Agriculture (CA):** Conservation agriculture refers the concept of (1) causing minimum disturbance to the soil surface by using no- or minimum-tillage; (2) keeping the soil surface covered all the time through practices such as retention of crop residue, mulching, or growing cover crops; (3) adopting crop sequences or rotations that include agroforestry in spatial and temporal scales; and (4) controlled traffic [FAO 2010]. Collectively these practices lead to an increase in water stable aggregates, greater SOC concentrations, and protection from wind and water erosion. Conservation agriculture-based crop management technologies include zero tillage (ZT) with residue recycling; laser assisted precision land leveling, direct drilling into the residues and direct seeding.

h) **Integrated Nutrient Management:** Integrated nutrient management, *i.e.*, the application of NPK mineral fertilizers along with organic manure, increases crop productivity, improves SOC content, and decreases soil loss and runoff.

i) **Tillage Practice:** Reduced tillage or zero tillage favors the better soil structure, infiltration and increase in pore space of soil. This helps in reducing the erosion. Bhardwaj 1998 shown that runoff and soil losses are greatly reduced with low intensity tillage (Table.7.7).

Table 7.7: Effect of tillage and mulch on runoff and soil loss

Conservation measures	Runoff (%)	Soil loss (tones/ha)
Conventional maize	61.2	15.76
Zero tillage without live mulch	41.8	12.00
Normal tillage with live mulch	31.2	7.00
Zero tillage with live mulch	21.6	3.30

j) **Riparian Strips:** These are merely buffer strips of grass, shrubbery, plants, and other vegetation that grow on the banks of rivers and streams and areas with water conservation problems. The strips slow runoff and catch sediment. In shallow water flow, they can reduce sediment and the nutrients and herbicides attached to it by 30% to 50%.

k) **Sloping Agriculture Land Technology (SALT):** SALT is a technology package of soil conservation and food production that integrates several soil conservation measures (Tacio 1993). Basically, SALT utilizes nitrogen-fixing trees as soil binder, fertilizer generator, and livestock feed source. The system also includes annual and perennial diversified food crops grown in the spaces between the hedgerows. SALT to help control soil erosion and increase crop yields.

Revegetation techniques

a) **Top Soiling:** The placement of topsoil over prepared subsoil prior to the establishment of vegetation to provide a suitable soil medium for vegetative growth while providing some limited short term erosion control capability.

b) **Temporary and Permanent Seeding:** The planting and establishment of quick growing and/or perennial vegetation to provide temporary and/or permanent stabilization on exposed areas. Temporary seeding is designed to stabilize the soil and to protect disturbed areas until permanent vegetation or other erosion control measures can be established.

c) **Hydroseeding:** Hydroseeding is a planting process that uses slurry of seed and mulch. It is often used as an erosion control technique. The application of seed, fertilizer and a paper or wood pulp with water in the form of a slurry, which is sprayed over the area to be brought under vegetation. It helps in establishing vegetation quickly while provides instant protection from impact of rain drops.

d) **Turfing:** A surface layer of earth containing a dense growth of grasses and its roots; (sod). Turfing is an artificial substitute for such a grassy layer, as on a playing field. The establishment and permanent stabilization of disturbed areas by laying a continuous cover of grass turf. To provide immediate vegetative cover to stabilize soil on disturbed areas.

Reclamation of gullies

- Reclamation of small gullies can be done through leveling and safe disposal of water, by Protection from biotic interference, peripheral bunding, safe disposal of runoff, terracing and gully plugs with plantation of locally adapted tree, shrub and grass species, wire meshes, sand bags, boulders, live hedges (vetiver) and etc.
- Clearing and leveling the bed, sodding, constructing a series of composite earth and brick masonry, check dams at 1.2 m vertical interval, and terracing the side slopes can reclaim a medium gully.
- The terrace faces, grassed outlets and earth check dams are stabilized by growing suitable grasses such as *Cenchrus ciliaris* and *Dichanthium annulatum*.
- For deep ravines the best land use is to put them under permanent vegetation, comprising locally adapted grass, shrub and tree species.

Soil conservation methods

i. Expansion of vegetative cover and protective afforestation
ii. Controlled grazing
iii. Flood control
iv. Prohibition of shifting cultivation
v. Proper land utilization
vi. Maintenance of soil fertility
vii. Land reforms, reclamation of wasteland
viii. Establishment of soil research institute and training of soil scientists and
ix. Effective agencies for soil management

Soil compaction

In India, about 90 M ha land is affected by various types of soil physical constraints (Table 7.8). These soils (both irrigated and rainfed) produce very low crop yields and have lower fertilizer nutrient use efficiency. The major soil physical constraints in Indian soils include low water retention and high transmission, slow permeability, surface and sub-surface mechanical impedance and shallow depth of the soils. These characteristics either restrict crop growth or reduce efficiency of basic inputs, such as water and fertilizer nutrients. Soil compaction is emerging as a serious problem affecting the yield of field crops leading to soil degradation worldwide. Compaction-induced soil degradation affects about 68 million hectares of land globally (Flowers and Lal, 1998). Soil compaction is one of the serious and unnecessary forms of soil physical degradation that can result in increased soil erosion and decreased in crop production.

Table 7.8: Area and states affected by various soil physical constraints in India

Physical constraints	Area (M ha)	Main states affected
Shallow depth	26.40	Andhra Pradesh, Maharashtra, West Bengal, Kerala and Gujarat
Soil hardening	21.57	Andhra Pradesh, Maharashtra and Bihar
High permeability	13.75	Rajasthan, West Bengal, Gujarat, Punjab and Tamil Nadu
Subsurface hard pan	11.31	Maharashtra, Punjab, Bihar, Rajasthan, West Bengal and Tamil Nadu
Surface crusting	10.25	Haryana, Punjab, West Bengal, Orissa and Gujarat
Temporary water logging	6.24	Madhya Pradesh, Maharashtra, Punjab, Gujarat, Kerala and Orissa

Source: Katyal, J.C., 2016

Compaction of soil is the compression of soil particles into a smaller volume, which reduces the size of pore space available for air and water. Most soils are composed of about 50 per cent solids (sand, silt, clay and organic matter) and about 50 per cent pore spaces. Soil compaction occurs when a force compresses the larger soil pores and reduces the air volume of the soil (Fig. 7.4). The continuity of the pores from the surface of the soil to deeper depths is disrupted, and thus, the transmission of water through soil and gases between the crop's roots and atmosphere is reduced (Fig. 7.5).

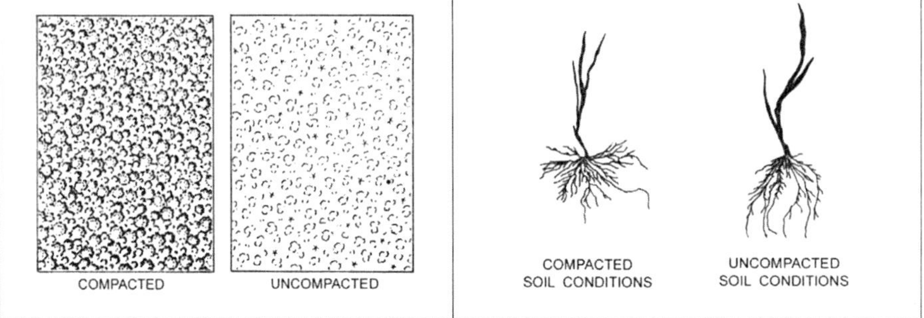

Fig.7.4: The effect of compaction on pore space

Fig. 7.5: The effect of compaction on crop root growth

Factors influencing soil compaction

The compaction of soil depends on its physical properties, water content and the nature of the force applied. The physical properties influencing compaction include the original bulk density, structure, texture and organic matter content in soil. In general, soils with fine textures (silt and clay), having low organic matter contents, high porosity and weakly aggregated structure are more susceptible to serious compaction. From a soil management stand point, all soils should be considered as capable of being compacted. The two major management factors that influence the soil's ability to compact are the soil moisture content at the time of any field operation and the contact pressure exerted by the implement or vehicle involved in the operation. Thus the extent of the soil compaction problem is a function of soil type and water content, vehicle weight, speed, ground contact pressure and number of passes, and their interactions with cropping frequency and farming practices (Larson *et al.,* 1994; Chamen *et al.,* 2003).

Effect of soil compaction

Soil compaction can have a number of negative effects on soil quality and crop production including the following: Soil compaction causes decrease in soil pore spaces, reduces infiltration rate of soil, decreases the rate of water penetration into the soil root zone and subsoil, increases the potential for surface water ponding, water runoff, surface soil waterlogging and soil erosion, reduces the ability of a soil to hold water and air, which are necessary for plant root growth and function. Soil compaction affects crop emergence as a result reduction in yield or yield potential of crops is observed. The inability of roots to penetrate compacted soil layers will result in less root penetration into the soil, root mass is reduced and a plant's ability to take up nutrients is reduced. As a result of surface soil layers having lower water storage capacities, plant roots remain closer to the surface and are therefore more susceptible to drought. Compacted soils often have higher subsurface soil moisture contents because soil water is unable to drain away freely and air movement in the soil is restricted. This reduction in internal drainage generally leads to surface ponding which may drown the crop.

Types of soil compaction: Soil compaction can occur at the soil surface in the form of soil crusting, or it can occur in the subsoil in the form of hardpan and other form is wheel traffic induced compaction.

1 Surface soil crusting: The compaction of soil caused due to effect of soil tillage and raindrop or irrigation water, result in soil crust formation. Surface crusting in red soil is due to the presence of colloidal oxides of iron and aluminium in soils which binds the soil particles under wet regimes. On drying it forms a hard mass on the surface. It is predominant in Alfisols but occurs in other soils too.

Causes: Soil tillage can bury much of the protective residue cover on the soil surface and degrade the granular structure of surface soils (mechanical crushing or breaking of larger soil aggregates). The impact energy of raindrops or irrigation droplets can also cause considerable degradation and breakdown of soil aggregates, the resultant soil particles is suspended in water and flow together along the runoff and then dry into a hard surface soil crust. The crusted soil can restrict water infiltration into soil and accelerates surface runoff. It restricts germination of seeds and retards root growth, Creates poor aeration in the rhizosphere and affects nodule formation in leguminous crops.

Management options to reduce soil compaction

- A short-term solution to soil crusting after seeding might be a light harrowing or rolling with packers to gently fracture the soil crust after seeding, to aid in seedling emergence through the crust.
- The best way to prevent soil crusting in fields is to both minimize tillage operations and ensure that a protective layer of residue remains on the soil surface to absorb the impact of water droplets before they strike

and break down stable soil aggregates. This can be achieved by reduced tillage or, preferably, by using organic mulches. These methods leave greater amounts of residue on the soil surface to reduce soil crusting and increase soil organic matter levels, leading to improved surface soil structure.

- Using crop management practices such as including a forage in the crop rotation, resistant crops to be grown, use of organic manures to increase the levels of soil organic matter which will aid in the development of a good granular-structured soil that has greater resistance to breakdown. In irrigated fields, it is also very important that water application is managed to ensure the infiltration rate of soil is not exceeded.

2. Subsurface compaction (hardpan/ tillage pan)

A tillage-induced compaction layer is sometimes called "hardpan," or "tillage pan" or "plow pan" and occurs in the layer of soil just below the depth of tillage. It occurs when soils are cultivated repeatedly at the same depth. The force of the tillage equipment, such as discs or cultivator shovels, can cause compression of the soil and smearing at the base of contact between the soil and tillage implement. Usually the compacted hardpan layer is one or two inches thick. Compaction will increase when soil moisture conditions are wet at the time of tillage and/or if soils have a higher silt and clay content. However, with coarser textured soils, the hardpan tends to be weaker and more friable, and may have less effect on crop production. In red soil sub surface hard pan is formed due to the illuviation of clay to the sub soil horizons coupled with cementing action of oxides of iron, aluminium and calcium carbonate. The sub soil hard pan is characterized by high bulk density(>1.8 Mg m-3) which in turn lowers infiltration, water holding capacity, available water and movement of air and nutrients with concomitant effect on the yield of crops.

Management options to reduce soil compaction

- It has generally been assumed that compacted soil layers break down naturally with annual freeze-thaw and wetting-drying cycles. When a hardpan soil layer persists, the soil may need some form of tillage to physically break up the hardpan. Tillage may reduce the hardpan compaction, but it may not address the sources of the problem. It may take three to five years of reduced tillage or zero tillage practices.
- To avoid the development of a tillage-induced hardpan, land should be direct seeded to minimize tillage. If soil is to be tilled, ensure that soils are not too moist, and vary the depth and direction of tillage.

- Chiselling technology to overcome the sub soil hard pan : The field is to be ploughed with chisel plough, a tractor drawn heavy iron plough at 50 cm interval in both the directions. Chiselling helps to break the hard pan in the sub soil, besides it ploughs up to 45 cm depth.
- Use of organic residues: maintaining the crop residue on the soil surface will help in prevent soil erosion and avoidthe intermix subsoil with topsoil.
- Inclusion of deep-rooted crops in crop rotation: Deep root crops/ taproot and fibrous-rooted crops should be included in cropping system, it could help to naturally alleviate the negative effects of compacted layer formed due to compaction.
- Addition of organic matter: Farm yard manure or press mud or composted coir pith should be used regularly and spread evenly on the surface. The field should be ploughed with country plough twice for incorporating the added manures. The broken hard pan and incorporation of manures make the soil to conserve more moisture.

3. Wheel traffic-induced compaction

Heavy farm equipment, including tractors, grain carts, combine harvester, trucks, can exert considerable weight onto the soil surface and into the subsoil. The effect of equipment weight can penetrate and cause compaction to a depth of about 24 inches when soils are very moist. Wheel traffic compaction has increased in the past several decades due to the increasing size of farm equipment and its repeated use to complete farm operations from sowing to harvest of crops. In rainy season, when soil moisture conditions tend to be higher, the risk of soil compaction is often the greatest.

Management options to reduce soil compaction

- Wheel traffic-induced compaction can be managed using good agronomic practices, deep tillage or both.
- Where wheel traffic has created compaction to a depth of 12 to 24 inches, use deep tillage to break the compacted soil zone.
- Limit traffic to 20-30% of field area by designating traffic lanes: It could save 70-80% area from traffic induced soil compaction.
- Limit machinery weight: Use of light weighted farm machinery or lower axle load lead to lesser compression of soil.
- Avoid working wet soil and improve field drainage: Never till the soil under wet condition, as under wet conditions soil are more prone to

compaction because of reduced soil aggregate stability and lubrication of particles due to higher wetness.

- Include deep-rooted crops in crop rotation: Deep root crops should be included in cropping system, it could help in naturally alleviate the negative effects of compacted layer formed due to compaction.
- Vary the depth of tillage or chiseling the soil: It could help in breaking the developed compacted layer.
- Addition of organic matter: It improves aggregates stability and soil structure, thus protect the soil against soil compaction.

Conclusion

Soil erosion in India is result of physical and anthropogenic activities. As soil erodes, it loses nutrients, reduces the crop yield, clogs water resources, and eventually turns the fertile area into desert. Soil erosion has serious on site and off site impacts for the whole economy. Erosion can be controlled by using the right means, tools, and methods at the right time at a right place. The most natural and effective way to prevent erosion control is through vegetation. Both agronomical and mechanical measures help in management of eroded soils.

Soil compaction increases bulk density and penetration resistance, decreases porosity, infiltration rate and hydraulic conductivity. The formation of compact layer at any depth resists the penetration of roots, its growth and development and also strongly affects soil-plant-water relations. The conservation agriculture should be practiced to reduce traffic on the soil and subsoiling/chiseling to remove hardpan developed due to traffic and puddling.

References

Balasubramanian, A. 2017. Methods of Controlling Soil Erosion. *Technical Report*. DOI: 10.13140/RG.2.2.22542.97609

Bhardwaj, S.P. and Sindhwal, N.S. 1998. Zero tillage and weed mulch for erosion control on slopping farm land in Doon valley. *Indian J. Soil Cons*., 26 (2): 81-85.

Bhattacharyya, R., Ghosh, B.N., Mishra, P.K., Mandal, B. Rao, C.S, Sarkar, D., Das, K. Kokkuvayil, S. A., Lalitha, M., Hati, K. M. and Franzluebbers, A. J. 2015. Soil Degradation in India: Challenges and Potential Solutions *Sustainability*, 7: 3528-3570. doi:10.3390/su7043528

Chamen, T., Alakukku, L., Pires, S., Sommer, C., Spoor, G., Tijink, F. and Weisskoff, P. 2003. Prevention strategies for field traffic induced subsoil compaction: a review. Part 2. Equipment and field practices. *Soil and Tillage Research,* 73: 161-74.

CSWCR&TI Vision, 2030. Vision 2030 of the Central Soil and water Conservation Research and Training Institute. 2011, Allied publisher, Dehradun, India, pp. 1–46.

Dhruvanarayan, V.V.N. and Ram, B. 1983. Estimation of soil erosion in India. *J. Irrig. Drain. Eng.,* 109: 419–434.

Flowers, M.D. and Lal, R. 1998. Axle load and tillage effects on soil physical properties and soybean grain yield on a mollic ochraqualf in northwest Ohio. *Soil and Tillage Research,* 48: 21-35.

Food and Agriculture Organization (FAO), 2010. *"Climate-Smart" Agriculture. Policies, Practices and Financing for Food Security, Adaptation and Mitigation*; FAO: Rome, Italy, p. 41.

Indian Society of Soil Science, 2002. Fundamentals of Soil Science. ISSS, New Delhi.

Judson, S. 1965. Physical Geology. Prentice Hall. NJ, USA. 3rd Edition, pp. 143-144.

Katyal, J.C., Chaudhari, S.K., Dwivedi, B.S., Biswas, D.R., Rattan, R.K. and Majumdar, K. 2016. Soil Health: Concept, Status and Monitoring. Bulletin of the Indian Society of Soil Science 30: 1-98.

Kaur, B., Gupta, S.R. and Singh, G. 2000. Soil carbon, microbial activity and nitrogen availability in agroforestry systems on moderately alkaline soils in Northern India. *Appl. Soil Ecol.* 15: 283–294.

Larson, W.E., Eynard, A., Hadas, A. and Lipiec, J. 1994. Control and avoidance of soil compaction in practice. In: *Soil Compaction in Crop Production,* edited by Soane BD and van Ouwerkerk C (Elsevier Science) Amsterdam, pp. 597-625.

Mandal, D., Sharda, V.N. and Tripathi, K.P. 2010. Relative efficacy of two biophysical approaches to assess soil loss tolerance for Doon Valley soils of India. *J. Soil Water Conserv.*, 65: 42–49.

Mandal, K.G., Hati, K.M., Misra, A.K., Bandyopadhyay, K.K. and Tripathy, A.K. 2013. Land surface modification and crop diversification for enhancing productivity of a Vertisol. *Int. J. Plant Prod.*, 7: 455–472.

Merritt, W.S., Letcher, R.A. and Jakeman, A.J. 2003. A review of erosion and sediment transport models. *Environmental modeling and software*, 18: 761–799.

Mishra, A., Sharma, S.D. and Khan, G.H., 2003. Improvement in physical and chemical properties of sodic soil by 3, 6, and 9 year-old plantations of *Eucalyptus tereticornis* bio-rejuvenation of soil. *For. Ecol. Manag.*, 184: 115–124.

Nair, P.K.R. 1993. An Introduction to Agro Forestry; Kluwer Academic Publishers: Dordrecht, The Netherland.

Narayana, V.V.D. and Babu, R. 1983. Estimation of soil loss in India. *Journal of irrigationand drainage engineering*, 109(4): 419–433.

National Bureau of Soil Survey & Land Use Planning (NBSS&LUP), 2004. *Soil Map (1:1 Million Scale)*; NBSS&LUP: Nagpur, India.

Sehgal, J. and Abrol, I.P. 1994. Soil degradation in India: status and impact, Oxford and IBH, New Delhi.

Singh, G. Ram Babu, Narain, P., Bhusan, L.S. and Abrol, I.P. 1990. Soil erosion rates in India□,Journal of Soil and Water Conservation. 47(1): 97–99.

Singh, G., Babu, R., Bhushan, N. and Abrol, I. P. 1990. Soil erosion rates in India. *Journal of Soil and Water Conservation*, 47(1): 97–99

State of the Environment India, 2001. Land degradation, Part III http://envfor.nic.in/sites/default/ files/soer/2001/soer.html. Accessed February 2014.

Tacio, H. D. 1993. Sloping Agricultural Land Technology (SALT): a sustainable agroforestry scheme for the uplands. *Agroforestry Systems,* 22(2): 145–152.

Ul Zaman, M., Bhat, S., Sharma, S. and Bhat, O. 2018. Methods to Control Soil Erosion-A Review. *Int. J. Pure App. Biosci.* 6(2): 1114-1121.

Internet/ web References

http://ecoursesonline.iasri.res.in/mod/page/view.php?id=125087

http://www.agritech.tnau.ac.in/pdf/3.pdf

http://www.ciesin.columbia.edu/docs/002-413/fig1.gif

http://www.weru.ksu.edu/new_weru/images/CreepSaltSusp.jpg.

https://link.springer.com/content/pdf/10.1007%2F978-3-319-19168-3_15.pdf

8

Water Logged Soils and Their Reclamation and Management

Asha Kumari Sinha

Submerged soils/Flooded soil

Submerged soils are the type of soils that are saturated with water for a sufficiently long time in a year. This gives the soil the following distinctive gley horizons resulting from oxidation-reduction processes:

i. Partially oxidized 'A' horizon high in organic matter.

ii. Mottled zone in which oxidation and reduction alternate.

iii. Permanently reduced zone which is bluish green in colour.

The soil is intermittently saturated with water, oxidation of organic matter is slow and it accumulated within the "A" horizon. Fe and Mn are deposited as rusty mottles or streaks if the diffusion of O_2 into the soil is slow in the second horizon, whereas if the diffusion is rapid, they are deposited as concretions.

Types of submerged soils

1. **Waterlogged soils-** Waterlogged soils are saturated with water for a sufficiently long time annually to give the soil the distinctive gley horizons resulting from oxidation-reduction processes. Saturation with water may be due to impermeability of the soil material, or the presence of an impervious layer, or a high water table.

2. **Marsh Soils-** These types of soils are more or less permanently saturated or submerged. Freshwater marshes can be classified according to their origin into upland, lowland, and transitional. Upland marshes are poor in bases and have pH 3.5-4.5 because they receive mainly rainwater. Lowland marshes are saturated or submerged with water-carrying bases and have pH 5.0-6.0.

3. **Paddy Soils-** Paddy soils are managed in a special way for the wet cultivation of rice. The management practices include:
    - Levelling of the land and construction of levees to impound water
    - Puddling (ploughing and harrowing the water-saturated soil)
    - Maintenance of 5-10 cm of standing water during the 4-5 months the crop is on the land
    - Draining and drying the fields at harvest
    - Reflooding after an interval which varies from a few weeks to as long as 8 months

These operations and oxygen secretion by rice roots results in the development of certain features peculiar to paddy soils.During the period of submergence, the soil undergoes reduction and turns into dark groy colour. Iron, manganese, silica and phosphate become more soluble and diffuse to the surface soil and move by diffusion and mass flow to the roots and to the subsoil. The cations displaced from exchange sites particularly by iron and migrate out of the reduced zone and are lost. When the soil is drained and dried, the reduced iron is re-oxidized and precipitated and leaving H^+ ions as the only major cation. Thus the soil is acidified and the clay disintegrates.

4. Subaqueous Soils- Subaqueous soils are formed from river, lake, and ocean sediments under continuous water columns.

Properties of submerged soil

A. Physical Properties of Submerged Soil

1. Depletion of oxygen
2. Accumulation of carbon dioxide
3. Swelling of colloids
4. Compaction
5. Destruction of aggregates
6. Bulk density high
7. Reduction of permeability (percolation rate)
8. Soil temperature decrease

B. Electro-chemical Properties of Submerged Soil

i. Redox potential(Eh)

ii. Soil pH

iii. Specific conductance

C. Chemical Properties of Submerged Soil

1. Soil reduction and transformation of different nutrient elements etc

D. Biological properties of Submerged Soil:

1. Change in soil biology

A. Physical properties of submerged soil

1. Depletion of oxygen

When a soil is submerged, water fills the pore spaces and replaces the air and oxygen supply of the submerged soil is virtually cut off. Oxygen can enter the soil only through molecular diffusion in the interstitial water. This process is 10,000 times slower than in gas-filled pores. Thus when a soil reaches saturation stage by water, the oxygen diffusion rate suddenly decreases. Most soil layers are nearly oxygen-free within a few hours after submergence except in a thin layer at the soil surface, and sometimes a layer below the plough sole.

The oxygen-diffusion in the water layer above the soil is very slow and the rate of O_2 consumption in reduced soil is high. Within a few hours of flooding, microorganisms use up the oxygen present in the water or trapped in the soil and render a submerged soil practically devoid of molecular oxygen. Because of this high demand of O_2 in submerged soil and slow O_2 supply through water, the soil is practically devoid of oxygen. This rapid depletion of O_2 takes place within a day or so of submergence. Some O_2 trapped in blocked pore spaces is rapidly utilized by facultative anaerobic organisms. Thus facultative anaerobic conditions predominate after some time.

In a submerged soil, the greater potential consumption of O_2 as compared to the available supply through the flood water results in formation of two distinctly different layers: An oxidized or aerobic layer and Reduced or anaerobic layer.

An oxidized or aerobic layer near soil surface:-In this layer O_2 is available through diffusion. Aerobic layer is formed at soil surface at soil-water interface, where oxidised situation prevails. The thickness of the aerobic (oxidised) surface layer is determined by the ratio of O_2 supply from the atmosphere to the O_2 consumption in the soil.

If the demand of O_2 is higher than the supply, then a thin oxidised surface layer (even one millimetre thick) is formed. A soil rich in organic matter or readily decomposable organic residues has a thin oxidized layer and if a soil which is poor in organic matter, the oxidised layer is thick. The thickness of oxidised layer may be upto several cm thick. In laterites, where percolation rates are high, the oxidized layer may be also several cm thick.

Reduced or anaerobic layer below soil surface:-Anaerobic layer is below soil surface, where no O_2 is available and reduced condition prevails.

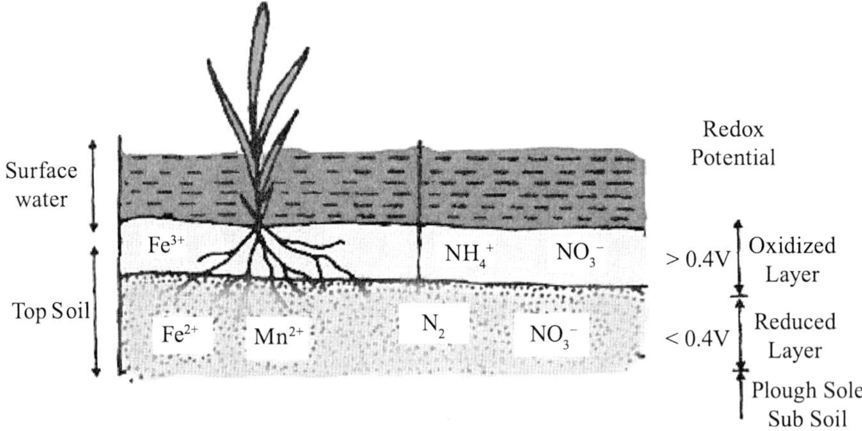

Fig. 8.1: Soil profile of a submerged soil

2. Accumulation of carbon dioxide: Submergence not only devoid the oxygen of soil but also drastically cuts down the escape of soil gases. Carbon dioxide, methane, hydrogen, and nitrogen produced in the soil tend to accumulate, which can be observed by rising bubbles when pressure build up, and escape as bubbles. During the first three weeks submergence some soils may generate CO_2 upto 2.5 t ha^{-1}.The partial pressure of carbon dioxide in a soil increases after submergence and reaches a peak of 0.2 – 0.8 bars 1-3 weeks later. The partial pressures of CO_2 affects the solubility of Ca^{2+}, Mg^{2+}, Fe^{2+} and Mn^{2+} cations. Carbon dioxide injury to rice may occur on acid soils low in iron, in organic soils, and cold soils.

Partial pressure of carbon dioxide (pCO_2) in a reduced soil determines soil conditions like pH, Eh, solubility of cations and thereby affect the specific conductance and cation exchange reactions. The increase in partial pressure of CO_2 on submergence is due to the accumulation of CO_2 initially produced by aerobic respiration followed by anaerobic decomposition of organic matter. Hence, the pCO_2 is higher in soils which contain high organic matter. The pCO_2 in acid soils is higher than in neutral soils. This may be due to the presence of higher proportion of H_2CO_3 in acid soils than in neutral soils. After one to four weeks of submergence, pCO_2 is decreased. This may be due to-

i. Escape of CO2 from the soil,

ii. Diluting effect of CH_4 produced in the later period of organic matter decomposition

iii. Reduction of CO_2 leaching losses and removal of CO_2 as insoluble carbonates.

iv. The rapid decline in the concentration of CO_2 in soils having high iron and manganese.

3. Swelling of colloids

When a dry soil is waterlogged, soil colloids absorb water and swell. The rate of absorption of water and volume increase of mineral soils depend on the clay content, type of clay mineral, and the nature of the adsorbed cations. Swelling is usually complete in one to three days. Swelling of clay increases with clay content. That means higher the clay content the greater the swelling. The expanding-lattice type of clays (montmorillonite and beidellite) swell more than that of fixed-lattice type (kaolinite and halloysite). Sodium clays swell more in comparison to calcium and potassium clays. As the moisture content of a soil increases the cohesion of water films around soil particles causes them to stick together rendering the soil plastic. At this moisture content soils are easily puddled. At higher moisture contents (as in flooded soils), cohesion decreases rapidly and making tillage easy, but penetration increases and soil strength decreases. Thus, the use of heavy machinery impractical on flooded soils.

4. Compaction

As a moisture content increases and the thickness of the water films around soil particles increases, the cohesion among particles decreases allowing them to slide over each other and the soil can be compacted readily to a greater density. In compacted soil, bulk density, micro pore-spaces, thermal conductivity and diffusivity and nutrient mobility increases and macro pore-spaces, hydraulic conductivity and water intake rate decreases.

5. Bulk density

When soil is compacted to such a degree that all pore spaces (air pore spaces + water pore spaces) are filled with water, no air pore spaces are present and no soil water is expelled from the pore spaces, soil is saturated and its bulk density is maximum.

6. Destruction of soil aggregates

Swelling of colloids and dissolution of cementing agents, such as iron oxide, further decrease aggregate stability. Sodic soils show marked aggregate breakdown on flooding, whereas soils high in iron and aluminium oxides and organic matter suffer little aggregate destruction.

7. Reduction of permeability (percolation rate)

Flooding decreases percolation rate in soils of low permeability even without puddling. This has been attributed to dispersion of soil particles, aggregate destruction, swelling and logging of pores by microbial slime. In porous, non-swelling soils, flooding (by providing a greater head of water) increases percolation.

8. Soil temperature

The effects of water logging on soil temperature follow from three important thermal properties of water

i) Its high specific heat

ii) High latent heat of vaporization

iii) Higher thermal conductivity than soil material.

The high specific heat prevents violent temperature fluctuation. The high heat of vaporization tends to keep flooded soils cooler than dry soils. The cooling effect is used to reduce high temperature injury in hot locations while the stabilizing effect is used to stop low temperature injury during night. On submergence thermal conductivity as well as thermal diffusivity increases. Low soil temperatures retard mineralization of organic nitrogen and phosphorus and favour the accumulation of carbon dioxide, organic acids, and excess water-soluble iron in flooded soils.

B. Electro-Chemical Changes in Submerged Soils

The main electro-chemical changes in submerged soils are

1. Redox Potential (Eh) of Submerged Soil

Oxidation-reduction is a chemical reaction in which electrons are transferred from a donor to an acceptor. The electron donor loses electrons and increases its oxidation number or is oxidized; the acceptor gains electrons and decreases its oxidation number or is reduced. The source of electrons for biological reductions is organic matter. The driving force of a chemical reaction is the tendency of the free energy of the system to decrease until, at equilibrium, the sum of the free energies of the products equals that of the remaining reactants. In a reversible oxidation-reduction reaction, this force can be measured in calories or in volts. Redox reactions in soils are mainly controlled by microbial activity. Organisms use organic substances such as carbon sources as electron donors during respiration. Molecular oxygen acts as the preferred electron acceptor as long as it is available. Flooding the field for subsequent rice cultivation cuts off the oxygen supply from the atmosphere, the microbial activities switch from aerobic to facultative and to anaerobic fermentation of organic matter where alternative electron acceptors are used.

In the absence of oxygen, facultative and obligate anaerobes use NO_3^-, Mn^{+4}, Fe^{+3}, SO_4^{2-}dissimilation products of organic matter, CO_2, N_2, and even H^+ ions as electron acceptors in their respiration reducing NO_3^- to N_2, Mn^{+4} to Mn^{+2}, Fe^{+3} to Fe^{+2}, SO_4^{2-} to H_2S, CO_2 to CH_4, N_2 to NH_4, and H^+ to H_2. Redox Potential is a quantitative measure of the tendency of a given system to oxidize or reduce substances. It is positive (+ve) and high in strongly oxidising systems but negative (-ve) and low in strongly reducing systems. There is no neutral point as in pH. Any chemical reaction which involves the exchange of electrons will be influenced by redox potential (Eh). A typical redox potential for environmental conditions ranges between +400 mV and +700 mV is associated with free dissolved oxygen. Below +400 mV, the oxygen concentration will begin to diminish and submerged soil conditions will become increasingly more reduced (>–400 mV) shown in table 8.1. Redox potentials are affected by pH and temperature.

Table 8.1: Critical Eh values of important components in submerged soils

Chemical Property form	Oxidized	Reduced form in submerged soils	Redox potential (Eh) in mv
Oxygen	O_2	H_2O	+380 to + 320
Nitrogen	NO_3^-	N_2O, N_2, NH_4^+	+280 to +220
Manganese	Mn^{4+}	Mn^{2+}	+280 to +220
Iron	Fe^{3+}	Fe^{2+}	+180 to +150
Sulphur	SO_4^{2-}	S^{2-}	-120 to – 180
Carbon	CO_2	CH_4	– 200 to – 280

Table 8.2: Oxidation-reduction potential found in rice soils ranging from well-drained to submerged conditions

Soil water condition	Redox potential (mv)
Aerated or well-drained	+700 to + 500
Moderately reduced	+ 400 to +200
Reduced	+ 100 to – 100
Highly reduced	-100 to -300

Redox potential is one of the most important soil physicochemical properties used to characterise a submerged soil. The course, rate and magnitude of the decrease in redox potential in a soil on flooding depend on the kind and amount of organic matter, the nature and content of electron acceptors, temperature and the duration of water logging. The presence of organic matter whether native or added, pushes forward the drop in redox potential and nitrate prevents it. Several soil factors influence the change in redox potential. Soils high in NO_3^- (more than 275 ppm NO_3^-) have positive potentials for several weeks after submergence; soils low in organic matter (less than 1.5%) or high in manganese (more than 0.2%) maintain positive potentials even 6 months after submergence; soils low in active manganese and iron (sandy soils) with more than 3% organic matter attain redox potentials of -200 mV to -300 mV within 2 weeks of submergence. However, the fairly stable potentials after several weeks of submergence lie between -200 and -300 mV.

Temperature above and below 25°C retard the decrease in redox potential but the degree of retardation varies with the soil. The retardation of reduction is most pronounced in acid soils and hardly noticeable in neutral soils high in organic matter. The fairly stable potentials attained after about 12 weeks of submergence are not affected in the temperature range of 15 to 45°C.When a soil (aerobic) is submerged, a sharp drop in the potential will observe and within a few weeks negative (-ve) potentials are observed in most of the soils. The submerged soils may exhibit potentials as low as -300 to -400 mV depending upon the type of soil organic matter, Fe and Mn content, a pH and temperature.

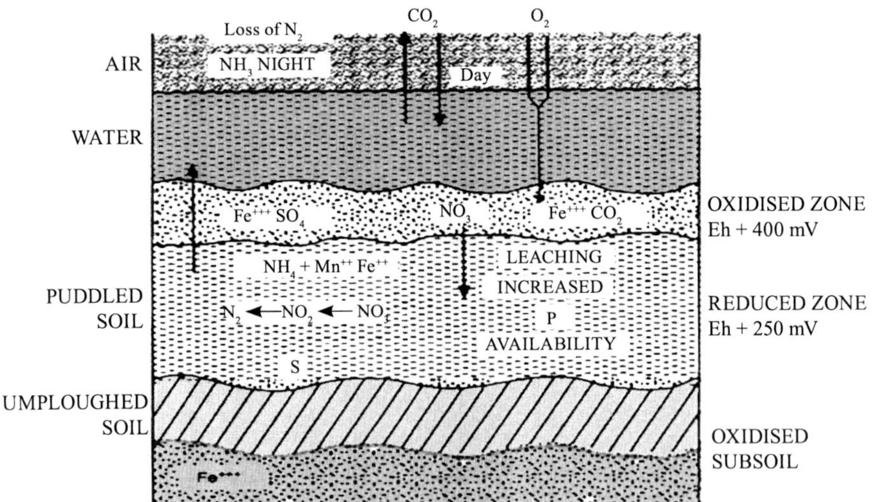

Fig. 8.2: Characteristics of oxidised and reduced zone in submerged soils

Sequential Reduction of Oxidation-Reduction Systems in Submerged Soils

After the disappearance of O_2 in submerged soils, the need for electron acceptors by facultative and obligate anaerobic micro-organisms results in the reduction of several oxidised soil components. In the presence of a readily available energy sources, microbes utilize several of the oxidised soil components and reduced the oxidation number of the oxidised atom, shown in fig 8.3.

Some of the oxidised soil components that undergo reduction after O_2 is depleted are reduced sequentially as follows in fig 8.4.

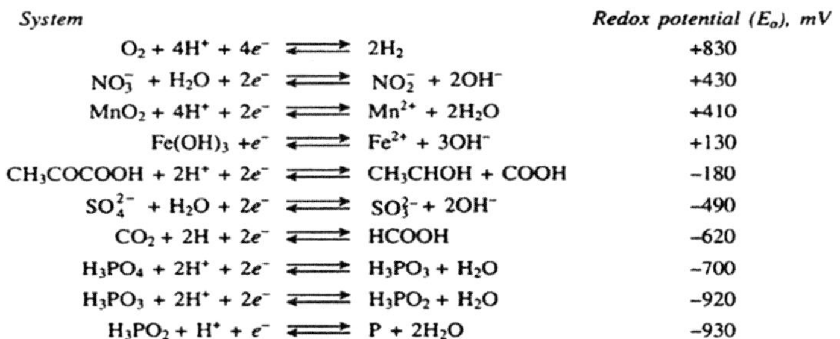

System		*Redox potential (E_o), mV*
$O_2 + 4H^+ + 4e^- \rightleftharpoons$	$2H_2$	+830
$NO_3^- + H_2O + 2e^- \rightleftharpoons$	$NO_2^- + 2OH^-$	+430
$MnO_2 + 4H^+ + 2e^- \rightleftharpoons$	$Mn^{2+} + 2H_2O$	+410
$Fe(OH)_3 + e^- \rightleftharpoons$	$Fe^{2+} + 3OH^-$	+130
$CH_3COCOOH + 2H^+ + 2e^- \rightleftharpoons$	$CH_3CHOH + COOH$	−180
$SO_4^{2-} + H_2O + 2e^- \rightleftharpoons$	$SO_3^{2-} + 2OH^-$	−490
$CO_2 + 2H + 2e^- \rightleftharpoons$	$HCOOH$	−620
$H_3PO_4 + 2H^+ + 2e^- \rightleftharpoons$	$H_3PO_3 + H_2O$	−700
$H_3PO_3 + 2H^+ + 2e^- \rightleftharpoons$	$H_3PO_2 + H_2O$	−920
$H_3PO_2 + H^+ + e^- \rightleftharpoons$	$P + 2H_2O$	−930

Fig. 8.3: General sequence of reduction of soil components

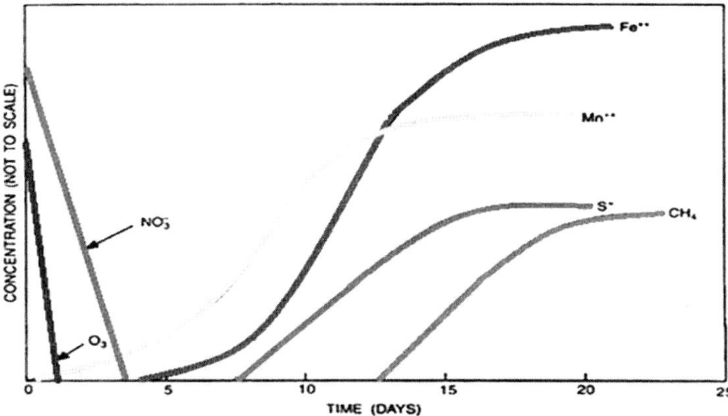

Fig. 8.4: The sequential reduction of various soil redox systems in submerged soils

ii. pH of submerged soil

After submergence the pH of acid soils increases and the pH of alkaline soils decreases. Under submerged condition, the pH of soils moves towards neutrality (pH 6.5-7.0). The increase in pH of acid soils may be due to the reduction of ferric (Fe^{+3}) and manganic (Mn^{+3}) ions and release of OH^- ions and also due to the presence of organic matter that enhance the process of soil reduction. Organic matter increases alkalinity. Since the reduction process removes hydrogen ions (H^+) from the soil solution, as reduction proceeds so the pH of the acid soil rises. Under submerged conditions, the pH of soils tend to be buffered around neutrality (pH6.5–7.0) due to the production of ferrous carbonate ($FeCO_3$) and ferrous hydroxide [$Fe(OH)_2$]. After submergence of alkali and calcareous soil, pH decreases. This may be due to the release of sufficient amount of carbon dioxide (CO_2) during decomposition of organic matter. This CO_2 reacts with soil water and produces hydrogen ions (H^+) and thereby decreases the soil pH.

$$\text{Organic matter} \xrightarrow{\text{Decomposition}} CO_2 \xrightarrow{H_2O} H_2CO_3 \xrightarrow{\text{Dissociation}} H^+ + HCO_3^-$$

iii. Specific conductivity

The specific conductivity under submerged conditions increases due to presence of Ca^{2+}, Mg^{2+}, Fe^{2+} (ferrous) ions in large amounts. Soils in which reduction proceeds at low speed, specific conductivity increases slowly to a level of 1-2 mmhos/cm and declines thereafter.The increase in conductance during the first few weeks of flooding is due to the release of Fe^{2+} and Mn^{2+} from the insoluble Fe^{3+} and Mn^{4+} oxide hydrates, the accumulation of NH_4^+, HCO_3^- and $RCOO^-$, and in calcareous soils, the dissolution of $CaCO_3$, $MgCO_3$ by CO_2 and organic acids. The kinetics of specific conductance varies widely with the soil characteristics. The strongly acidic lateritic soils show a steep increase in specific conductance during first week of submergence. Soils high in organic matter attain higher peaks.

C. Chemical properties of submerged soil

Change in soil chemistry

When organisms respire, they convert carbon compounds like glucose and fats into usable energy. During respiration, the carbon compounds release electrons. If these electrons are absorbed by oxygen molecules, the process is aerobic respiration. When they are absorbed by something other than oxygen, the process is anaerobic respiration. Anaerobic respiration happens where oxygen is depleting or absent such as submerged soils. Compounds that gain or lose electrons during respiration become new compounds with new properties. A compound losing an electron(s) during respiration is oxidized while a compound gaining an electron(s) is reduced. Both types of reactions happen simultaneously. When soil is submerged, organic matter in the soil is oxidized and soil components such as iron, manganese, and sulphur are reduced. These reactions happen in a predictable sequence. Compounds requiring the smallest amount of energy for reduction are reduced first. Those requiring more energy are going to be reduced when reduction of the first compound is all most complete. The reduction process will stop if there are not enough energy or electron acceptors to continue, shown in fig 8.5.

The compounds appearing inside the box below are common to aerated soil. The compounds in the soil change following submergence.

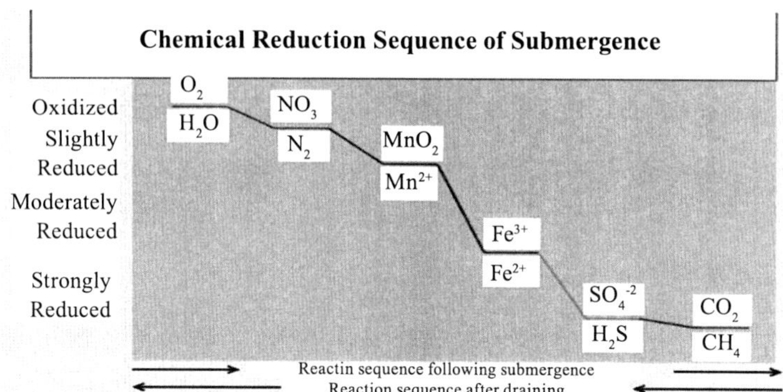

Fig. 8.5: Sequence of chemical reduction in submerged soils

The sequence of the compounds from left side to right side is the order in which they will be reduced. The reduction are going be starting with oxygen and ending with carbon dioxide. When soil is drained, oxygen begins to penetrate the soil and react with the reduced compounds. The reduced compounds are then oxidized in the reverse sequence starting on the right side and moving towards left side.

Other features change with soil chemistry

- Dominant form of elements in aerated versus submerged soil.
- Changes in chemical forms result in soil colour changes. While light colours indicate a well-drained soil, submerged soils are grey or blue-green colour
- Soil pH is another characteristic that changes because of the changes in soil chemistry. After flooding, all types of aerated soils, whether they are acidic or alkaline, moves towards neutral pH.

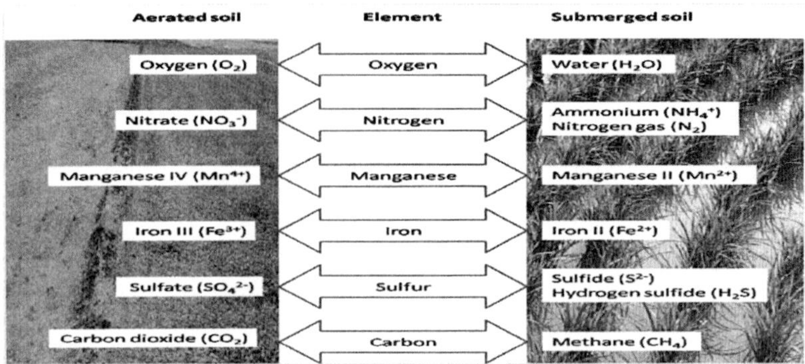

Fig. 8.6: Dominant form of elements in aerated and submerged soil

Transformation of important nutrient elements in submerged soils

1. Transformation of nitrogen in submerged soil

Transformation of nitrogen in aerobic layers of a submerged soil: In aerobic layers of a submerged soil the decomposition of organic matter proceed towards the production of NO_3^-.Organic form of nitrogen undergoes mineralization to NH_4^+, oxidation of NH_4^+ to NO_2^- and further oxidation of NO_2^- to NO_3^-.

In aerobic surface layer of soils-

$$\text{Organic forms of nitrogen} \xrightarrow[\text{Ammonification}]{\text{Mineralization}} NH_3^+ \xrightarrow{\text{Microbial oxidation}} NO_2^- \xrightarrow{\text{Microbial oxidation}} NO_3^-$$

In aerobic surface layer, conditions are similar to those of a well-drained soil. But in an anaerobic soil layer O_2 is absent and the absence of O_2 inhibits the activity of the Nitrosomonas micro-organisms (oxidising micro-organisms) which oxidises $NH4^+$. Due to inhibition of activity of oxidising micro-organisms, oxidation of ammonia stops. Therefore, nitrogen mineralization stops at the NH_4^+ form.

In anaerobic surface layer of soils-

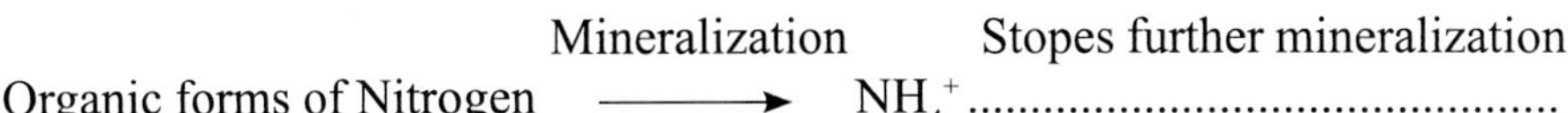

The accumulation of ammonia in submerged soils is, therefore, a good index of the capacity of a soil to meet up the demand for nitrogen to the rice crop. The transformation of nitrogen occurs in the aerobic and anaerobic layers of a submerged soil. In the aerobic surface layer, conditions are similar to those of a well-drained soil and nitrogen mineralization proceeds to the NO_3^- form. The presence of an aerobic layer above the anaerobic layer is the major reason of instability of nitrogen in submerged soils and results in considerable loss of nitrogen through nitrification-denitrification reactions.

Nitrate is stable and is not subject to denitrification as long as it remains in the surface aerobic layer, but it readily diffuses downward into the anaerobic layer and undergoes denitrification as a result of a gradient in the NO_3^- concentration between the aerobic layer and the anaerobic layer.This process can proceed as long as NO_3^- is formed in the aerobic layer, and that can readily happen if there is a source of NH_4^+ in the aerobic layer that can be nitrified (NO_3^-). The removal of NH_4^+ in that layer by nitrification creates a concentration gradient which causes NH_4^+ to diffuse upward form the anaerobic layer.

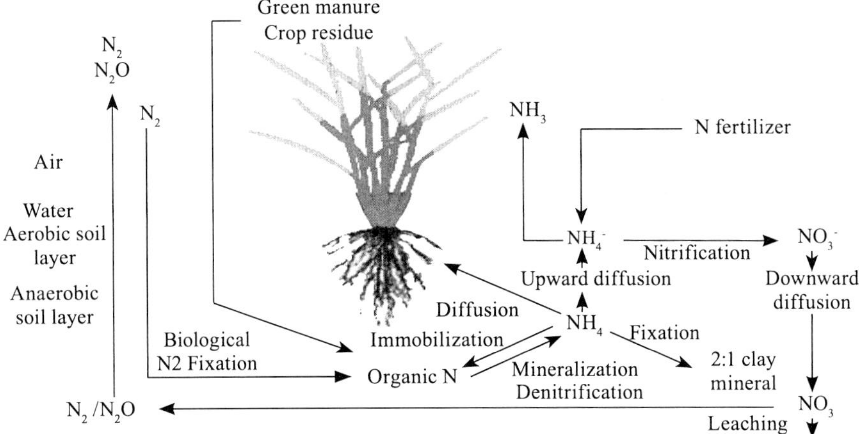

Fig. 8.7: Sequential conversions of organic nitrogen to elemental nitrogen in submerged soil

2. Transformation of phosphorus in flooded soils

Due to submergence, hydrolysis of aluminium phosphate and reduction of ferric phosphate releases phosphorus in available form. Besides, release of occluded phosphorus by reduction of the hydrated ferric hydroxide coating, organic anions displacing phosphorus from ferric and aluminium phosphate, enhancement in solubility of calcium phosphate due to decrease in pH caused by carbon dioxide accumulation in calcareous soils and increase in phosphorus availability due to mineralisation of organic residues, makes phosphorus more available under submerged conditions. Thus, on submergence, the availability of native as well as applied phosphorus increases in the soil.

Under alternate wetting (submergence) and drying, the transformations of P will not be similar to that of continuous submergence. Alternate wetting and drying reduces the availability of P in Al-P fraction and that increase in Fe-P fraction.

Under submerged conditions, due to reduction and chelation, green manuring and addition of organic residues increase the solubility and availability of phosphates in soil. However, inorganic phosphates may change to organic phosphorus under waterlogged conditions reducing availability of phosphorus.

Phosphorus may also be leached in soluble forms under submerged conditions.

The rise in phosphorus availability on submergence may be attributed to the following mechanisms

(i) Release of P from the mineralization of organic residues

(ii) Reduction of $FePO_4 . 2H_2O$ to the more soluble $Fe_3(PO_4)_2 . 8H_2O$ and increase in solubility of $FePO_4 . 2H_2O$ and $AlPO_4 . 2H_2O$ caused by the increase in pH coupled with the reduction of acid soils.

(iii) Release of co- precipitated or occluded phosphorus due to reduction of ferric oxy-hydroxide.

(iv) Displacement of P from ferric and aluminium phosphates by organic anions

(v) Increase in solubility of calcium phosphates ($CaHPO_4.2H_2O$, $Ca_4H(PO_4)_3.3H_2O$, $Ca_{10}(PO_4)_6(OH)_2$, $Ca_{10}(PO_4)_6CO_3$ and $Ca_{10}(PO_4)_6F_2$) associated with the decrease in pH caused by the liberation of CO_2 in the calcareous soils.

(vi) The release of P due to anion exchange reactions between clay and phosphate or organic anions and phosphate.

The decrease in the concentration of available P at the later period of submergence may be due to the fixation (through adsorption) of released phosphorus by clay colloids (kaolinite, montmorillonite and hydrous oxides of Fe and Al). The decreased concentration of phosphorus may also be due to the decreased solubility of phosphorus associated with calcium (Ca-P).

3. Transformation of potassium in flooded soils

Potassium is present in soils in four forms (Soluble K, Exchangeable K, Non-exchangeable K and Mineral K), which are in dynamic equilibrium. With flooding or water logging or submergence soluble ferrous (Fe^{2+}) and manganous (Mn^{2+}) ions increase and exchangeable K^+ is then displaced into the soil solution. The increase in soluble K^+ after water logging is closely related to the ferrous ion (Fe^{2+}) content of the soil solution.

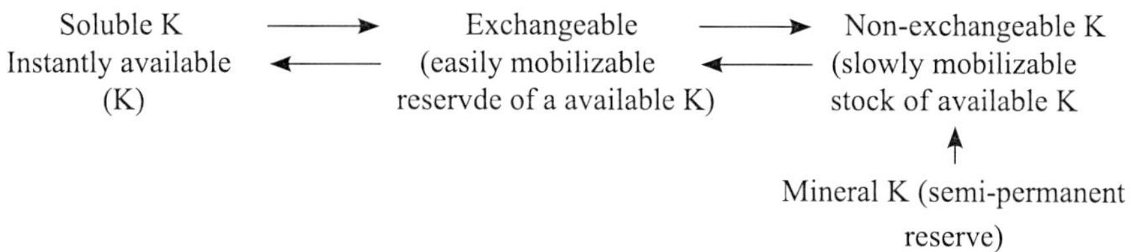

The release of K^+ from micas may be the contributing factor for the increase in K^+ in soil and that release depends on the various factors like tetrahedral rotation, degree of tetrahedral tilting, OH groups orientation, degree of K^+ depletion from the soil solution, hydronium ions (H_3O^+), biological activity and complexing organic acids, inorganic cations etc. Other than these, charge density and the configuration of the oxygen about exchange sites probably determines the release of K^+ and thus increases the concentration of K^+ in the soil solution.

It has been also reported that the availability of applied potassium decreases in waterlogged soils due to formation of Fe-K sparingly soluble complexes.

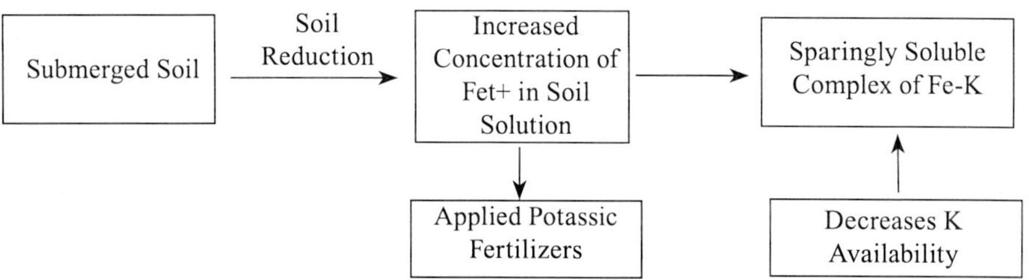

Fig. 8.8: Availability of applied potassium decreases in submerged soils

4. Transformation of sulphur in flooded soils

The main transformations of sulphur in flooded soils are –

- Reduction of sulphate (SO_4^{2-}) to sulphide (S^{2-})
- Dissimilation of the amino acids, cysteine, cystine and methionine to H_2S.
- Methyl-thiol has been found in submerged soils and the bad odour of putrefying blue-green bacteria in a reservoir has been attributed to dimethyl sulphide and methyl, butyl and isobutyl thiols.

The main product of transformations of the sulphur in submerged soils is hydrogen sulphide (H_2S) and it is derived largely from SO_4^{2-} reduction. In submerged soil H_2S (hydrogen sulphide) may react with various heavy metals (Zn, Cu, Cd, Pb etc.) to give their insoluble sulphides. Thus the availability of these metals may be reduced. Because Fe^{3+} reduction to Fe^{2+} precedes

SO_4^{2-} reduction, Fe^{2+} will be present in the soil solution at the time hydrogen sulphide (H_2S) is produced and that hydrogen sulphide will be converted into insoluble iron sulphide (FeS).

In submerged soil

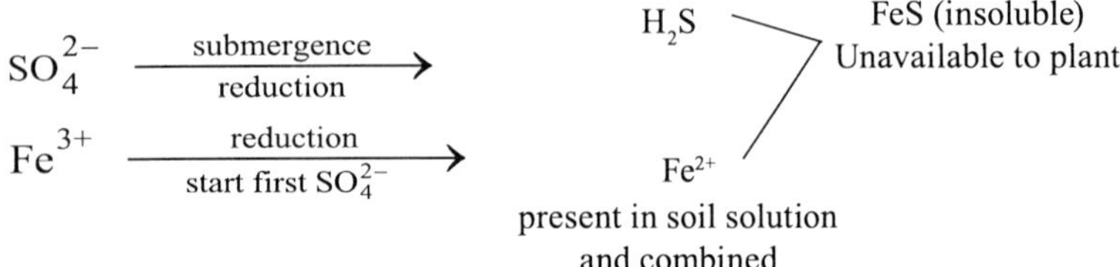

In muck and sandy soils low in iron, however, where iron is inactivated by the complex formation with organic matter, FeS formation may not take place. In that situation H_2S toxicity to the rice plant is possible. When an acid soil is flooded, the concentration of water soluble SO_4^{2-} increases initially and after that the concentration of the SO_4^{2-}decreases slowly. In flooded soil the initial increase in SO_4^{2-} concentration is due to the release (following increase in pH) of SO_4^{2-}, which is strongly sorbet at low pH by clay and hydrous oxides of Fe and Al.

The reduction of SO_4^{2-} has three implications

(i) Sulphur supply may become insufficient

(ii) Zinc and copper may be immobilized

(iii) H_2S toxicity may arise in soils low in iron

5. Transformation of Iron in flooded soils

Among the transformations in flooded soil, the reduction of iron with a concomitant increase in its solubility is the most important. When a soil is submerged the most important chemical change that takes place is the reduction of iron and the accompanying increase in its solubility. The intensity of reduction depends upon time of submergence, amount of organic matter, active iron, active manganese, nitrate etc. On submergence, due to reduction of Fe^{3+} to Fe^{2+}, the colour of soil changes from brown to grey and large amounts of Fe^{2-} enter into the soil solution. Acid soils rich in organic matter and active iron may attain high level of Fe^{2+}, in the initial stage of submergence which may show a steep roughly exponential decrease to levels of 50-100 ppm that persists for several months. Low temperature retards the peak and broadens the area under it. Soils high in organic matter but low in active iron give high concentrations of Fe^{2+}, which also persists for several months. In neutral and calcareous soils, the concentration of water-soluble iron rarely exceeds 20 ppm at any period of submergence. It is apparent that the concentration of ferrous iron (Fe^{2+}) increases initially to some peak value and after that decreases slowly with the period of soil submergence. Five to 50% of the free iron oxides present in a soil may be reduced within a few weeks of submergence, depending on the temperature, organic matter content and the stability of the oxides. Organic matter also enhances the rate of reduction of iron in submerged soils. Soils high in organic matter but low in active iron give high concentrations of ferrous ion(Fe^{2+}), which also persists for several months.

The initial increase in the concentration of ferrous iron (Fe^{2+}) on soil submergence is caused by the reduction that are

$$\underset{\text{(Insolute)}}{Fe(OH)3 + e} \xrightleftharpoons{\text{Reduction}} \underset{\text{(Soluble)}}{Fe^{2+}} + 3OH^-$$

The decrease in the concentration of Fe^{2+} following the peak rise is caused by the precipitation of Fe^{2+} as $FeC0_3$ in the early stages where high partial pressure of CO_2 prevails and as $Fe_3(OH)_8$ due to decrease in the partial pressure of $CO_2(pCO_2)$.

$$\underset{\text{(Soluble)}}{2Fe^{2+}\ 3CO_2^+ +} \longrightarrow \underset{\text{(Insoluble)}}{2FeCO_3}$$

The reduction of iron has some important consequences

(i) The concentration of water soluble iron increases

(ii) pH increases

(iii) Cations are displaced from exchange sites

(iv) The solubility of P and Si increases

(v) New minerals are formed

In neutral and calcareous soils, the concentration of water soluble iron rarely exceeds 20 ppm at any period of submergence.

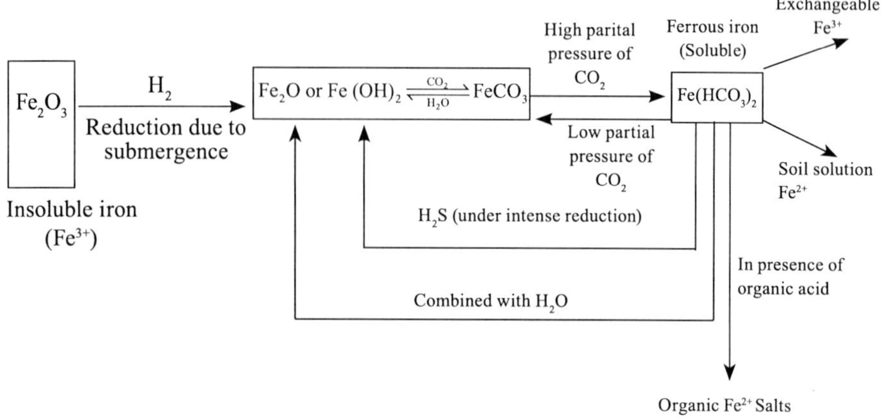

Fig. 8.9: A schematic representation for the transformation of iron in submerged soils

6. Transformation of Manganese in flooded soils

The main transformations of manganese in submerged soils are the reduction of manganic (Mn^{4+}) to more soluble manganous (Mn^{2+}) form and almost similar to that of iron transformation. Like iron, Manganese transformation in flooded soil is highly dependent on both pH and redox potential. In submerged soils, the transformation of Mn results an increase in the concentration of water soluble Mn^{2+}, precipitation of manganous carbonate ($MnCO_3$), and re-oxidation of Mn^{2+} diffusing or moving by mass flow to oxygenated interfaces in the soil. When an aerobic laterite soil is submerged the reduction of manganic manganese (Mn^{4+}) occurs almost concurrently with the nitrate (NO_3^-) reduction, but this reduction precedes that of Fe^{2+} reduction. The concentration of Mn^{2+} (water soluble) increases initially and after that declines with the period of soil submergence.

$$MnO_2 + 4H^+ + 2e^- \xrightleftharpoons{\text{reduction}} Mn^{2+} + 2H_2O$$

(water soluble)

Manganous ion, formed during soil reduction, may either remain in solution or be adsorbed on the exchange complex under acidic or slightly acidic conditions. The initial increase in the concentration of Mn^{2+} may be due to the reduction of soil Mn^{4+} and the decrease of the Mn^{2+} of the later period may be due to the precipitation of Mn^{2+} on manganous carbonate ($MnCO_3$) and manganous [$Mn(OH)_2$] in the soil solution. In flooded soils with near neutral pH, the reduced Mn^{2+} is precipitated as $MnCO_3$ or as oxides and hydroxides. Manganese in soil can exist in at least four different forms; water soluble, exchangeable, reducible and residual. Unlike iron which forms stable complexes, soluble manganese is present largely in the ionic form showed that manganese is converted from the reducible (oxidised form) to the exchangeable and water soluble forms during the course of soil reduction.

The kinetics of manganese reduction varies markedly from soil to soil. The changes in water soluble Mn^{2+} concentration depend upon the pH, organic matter content and active Mn content of the soils. The mobilization of Mn in soils is markedly increased after submergence due to the reduction of manganic compounds to more soluble forms as a consequence of the anaerobic metabolism of soil bacteria. Manganese reduction in flooded soil can be either chemical or microbiological. Microbiological reduction is likely to predominate at pH5.5 to 6,0 .Acid lateritic soils high in active Mn regardless of organic matter content will give higher peak of water soluble Mn^{2+} concentration sharply and low organic matter content delayed the peak. Strongly acid soils with relatively low in Mn content will also give lower peaks. The smallest peak will produce in slightly alkali soils and in soils very low in Mn content.The transformation of Mn in submerged soils largely depends on the oxidation-reduction reactions and the reduction of Mn^{4+}occurs when the redox potential value is within a range from +200 to +400 mV.

$$Mn^{4+} \xrightarrow[\text{submerged condition}]{\text{reduction}} Mn^{2+} \text{ (Occurs at a redox potential value of +200 to +400 mV)}$$

Organic matter affects the manganese transformation in soils through the following ways

i. The production of complexing agents that effectively decreases the activity of free iron in solution.

ii. The reduction in the oxidation-reduction potential of the soil either directly or indirectly through microbial activity.

iii. The stimulation of microbial activity that leads to the incorporation of Mn in biological tissue.

7. Transformation of zinc in flooded soils

Transformation of zinc in submerged soils is not involved in the oxidation-reduction process like that of iron and manganese. However, the reduction of hydrous oxides of iron and manganese, changes in soil pH, partial pCO_2, formation insoluble sulphide compound etc. In soil on submergence is likely to influence the solubility of Zn in soil either favourably or adversely and consequently the Zn nutrition of low or high.

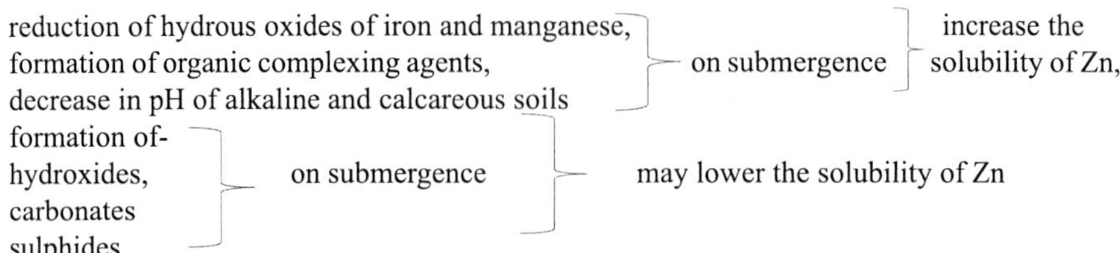

The reduction of hydrous oxides of iron and manganese, formation of organic complexing agents, and the decrease in pH of alkaline and calcareous soils on submergence are found to increase the solubility of Zn, whereas the formation of hydroxides, carbonates, sulphides may lower the solubility of Zn in submerged soils. Zinc deficiency in submerged rice soils is very common due to the combined effect of increased pH, HCO_3^- and S_2^- formation.

The solubility of native forms of Zn in soils is highly pH dependent and decreases by a factor of 10^2 for each unit increase in soil pH. When an aerobic soil is flooded, the availability of native and applied, both Zn decreases and the magnitude of such type of decrease vary with the soil properties. The transformation of Zn in soils is greatly influenced by the depth of submerged and application of organic matter. If an acid soil is submergence, the pH of the soil will increase and thereby the availability of Zn will decrease. On the other hand, if an alkali soil is submerged, the pH of the soil will decrease and as a result the solubility of Zn will generally increase.

Reasons for decrease in availability of Zn in submerged soil

The availability of Zn decreases due to submergence may be attributed to the following reasons:

(i) Formation of insoluble franklinite ($ZnFe_2O_4$) compound in submerged soils.

(ii) Formation of very insoluble compounds of Zn as ZnS under intense reducing conditions,

(iii) Formation of insoluble compounds of Zn as $ZnCO_3$ at the later stage of soil submergence owing to high partial pressure of CO_2(pCO_2) arising from the decomposition of organic matter,

(iv) Formation of $Zn(OH)_2$ at a relatively higher pH which redeuces the availability of

(v) Adsorption of soluble Zn^{2+} by oxide minerals e.g. sesquioxides, carbonates, soil organic matter and clay minerals etc. decreases the availability of Zn.

(vi) Formation of a variety of other insoluble zinc compounds which reduces the availability of Zn in submerged soil e.g. high phosphatic fertilizer induces the decreased availability of Zn^{2+}.

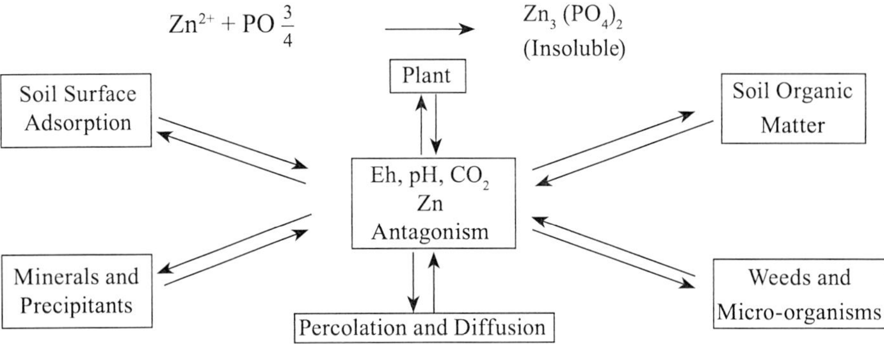

Fig. 8.10: Dynamic equilibrium of Zn in submerged soil

It portrays that rice receives Zn from the soil solution and the exchangeable and adsorbed solid phase including the soil organic fractions. The solubility of Zn might be controlled by Zinc sulphide (ZnS, Sphalerite) in the presence of traces of hydrogen sulphide (H_2S) in submerged soils. Zinc is stable in submerged soils. So it can be concluded that higher the pH and poorer the aeration, the higher is the likelihood of Zn deficiency if the soil solution Zn activity is controlled by sphalerite (ZnS). Therefore, many types of chemical reactions in soils influence the availability of Zn to rice. For instance, high manganese concentration antagonises Zn absorption and translocation. Calcium and magnesium may also affect Zn uptake.

The reversible pH change of the submerged soils, where the pH tends to increase in acid soil and decrease in alkaline soils, undoubtedly modify the Zn equilibrium concentration in the soil solution. Since the solubility of Zn minerals and Zn sorbed by soil colloids is pH dependent (higher at higher pH), an increase in the pH of an acid soil when submerged will tend to decrease the Zn concentration in the soil solution. However, in alkaline soil, initially Zn uptake increases as the pH decreases after submergence. Submerged alkali or calcareous soils possess all the essential characteristics for the formation of high amount of bicarbonate (HCO_3^-) ions which helps Zn^{2+} rendering unavailable to plants by forming insoluble $ZnCO_3$ compound. Along with these, the availability of Zn in submerged soils is governed by the mutual interaction of quantity (q) intensity (c), and kinetic parameters as regulated by the adsorption, desorption, chelation and diffusion of Zn from soils to the plant roots.

Different crop management factors combinedly influence the availability of Zn to rice like, native Zn content of the soil, soil pH, organic matter, submergence, partial pressure of CO_2, HCO_3^-, organic acids, various natural interactions, environmental effects and water quality, etc.

8. Transformation of copper in flooded soils

Most of the copper in soils is highly insoluble and can only be extracted by strong chemical treatments which dissolve various mineral structures of solubilized organic matter. Generally, the concentration of copper in soil solutions is very low.

pH values	Dominant species of Cu in soil solution
Below pH 6.9	Divalent Cu^{2+}
Above pH 6.9	$Cu(OH)_2^0$
pH 7.0.	$CuOH^+$

Hydrolysis reactions of copper ions are

$$Cu^{2+} + H_2O \rightleftharpoons CuOH^+ + H^+$$

$$CuOH^+ + H_2O \rightleftharpoons Cu(OH)_2^0 + H^+$$

Solubility of copper is very pH dependent and for each unit decrease in soil pH, it increases several times (approx. 100 times). The transformation of copper in submerged soils is not involved in oxidation-reduction reactions. The behaviour of copper in submerged soils is influenced by simple submergence in soils.

It is apparent that copper exists in soils as different discrete chemical pools which are as follows

(i) Water soluble plus exchangeable Cu.

(ii) Copper associated with clay minerals.

(iii) Organically bound Cu.

(iv) Copper associated with different oxides in soils.

(v) Residual copper.

The amount of each type of copper in soils depends on soil pH, amount of organic matter, clay content, oxides of Fe and Mn etc. All these above types of Cu are indynamic equilibrium in soils.In submerged soils, copper comes into the soil solution and becomes available to the plant.

$$\text{Soil } Cu^{2+} + H^+ \rightleftharpoons \underset{\text{(soluble)}}{Cu^{2+}}$$

$$Cu_2(OH)_2\,CO_3 + 4H^+ \rightleftharpoons Cu^{2+} + CO_2 + 3H_2O$$

(Malachite, and abundant
Source of Cu in
submerged soils)

When an acid soil is submerged, the soil pH increases and the release of copper decreases , whereas submerging an alkali and calcareous soils, the amount of copper in soil solution increases to a lesser degree. However, in most of the soils, submergence reduces the availability of copper and thereby creates deficiency to plants.

9. Transformation of Boron and Molybdenum in flooded soils-Since the solubility of the oxyanionic forms of these two elements is very much dependent on pH, organic matter content, clay content etc. The availability of Boron and Molybdenum in submerged soils is related to the changes in soil pH. Submerging an acid soil caused growth in the amount of available Mo content during the initial period, which remained almost unchanged at the later period. This increase might be due to the increase in soil pH and desorption of MoO_4^{2-} from oxides and hydroxides of Fe and Mn. It has been observed that the concentration of B in soil solution remains more or less constant after submergence.

Table 8.3: Change in organic matter and availability of plant nutrients in soils following their submergence under water

Chemical property	Change (s) following soil submergence
pH	Favours convergence to neutral pH
Organic matter	Favours accumulation of organic C and N
Ammonium-	N Release and accumulation of ammonium favoured
P	Improves P availability, especially in soils high in Fe and Al oxides
K	K improves through exchange of K
Ca, Mg, Na	Favours release of Ca, Mg and Na solution
S	Sulphate reduction may reduce sulphur availability
Fe	Iron availability improves in alkali and calcareous soils, but Fe toxicity may occur in acidic soils high in reducible Fe
Al	Al toxicity is generally absent, except perhaps in acid sulphate soils
Cu, Zn and Mo	Improves availability of Cu and Mo but not of Zn
Reduction products	Production of supplied and organic acids, especially in degraded soils may cause toxicity or injurious effects to growing plants.

D. Biological properties of submerged soil

Change in soil biology

Aerated soil contains tremendous diversity and quantity of organisms and microorganisms. Submergence starts a biological transition. When oxygen within the soil is depleted, aerobic organisms die or become dormant. They are replaced by two organisms surviving without oxygen called facultative and obligate anaerobes. The transition from aerobic to anaerobic organisms takes from a few hours to a few days. The majority of facultative and obligate anaerobes are specific types of bacteria. They use chemical compounds and other soil substrates oxygen for respiration, some protozoa and nematodes can also be present.

Causes of waterlogging

Main factors which help in raising the water table are:

i. The rate of percolation increases due to inadequate drainage of overland run-off and in turn helps in raising the water table.

ii. The water from rivers may penetrate into the soil.

iii. Seepage of water from water bodies like earthen canals also adds significant quantity of water to the underground reservoir continuously.

iv. Sometimes subsoil does not permit free flow of subsoil water which mayac celeratethe process of raising the water table.

v. If irrigation water is used in excess it may cause appreciable raise the water table.

Problems of water-logged soils are

- **Water depth:** Low land areas are usually flooded to depths of about 50 cm and hence limits the crop production.
- **Poor aeration:** Due to water-logging, the incoming water in the soil displaces a part of the soil air which then moves out into the atmosphere. Thus water logged soils are poorly aerated.
- **Soil structure:** Continuous water stagnation destroys the soil structure and makes the soil compact.
- **Soil temperature:** water-logging lower down the soil temperature. Specific heat of moist soil higher than dry soils.
- **Soil pH:** In waterlogged soil, pH moves towards neutral pH. That mins, pH tends to increase in acidic soils and decrease in alkaline soils.

- **Salinity**: salinity is an important constraint to rice production in many coastal lowlands as well as in alkaline soils.
- **Effect on crops**: Under water-logged condition all field crops (except rice) can not survive due to poor aeration and unavailability of nutrients to the plant.

Availability of nutrients

- **Nitrogen:** Due to lower temperature and reduced condition, mineralization of organic nitrogen is reduce and hence Nitrogen (N) deficiency is extremely common in water-logged soils.
- **Phosphorus:** The inorganic forms of P in flooded soils are usually present at higher levels than in upland soils.
- **Potassium:** Potassium response is apparent in many lowland soils. Flooding and puddling of the soils during lowland preparation may considerably increase the concentration of K in soil solution because of displacement of exchangeable K by the large amounts of iron and manganese in the soil solution.
- **Sulphur:** Sulphur deficiency has been reported from many lowland areas.
- **Zinc:** Zinc deficient is mostly observed in wetland alkaline soils. In general deficiency of zinc (Zn) to rice crop in wetland conditions is reported.
- **Iron and Manganese:** In submerged soil, iron and manganese are available in excess which causes toxicity to the plant

Severity of waterlogging

Waterlogging throughout the year and at different soil depths can be integrated by the Sum of Excess Water which occurs each day in the primary root zone of the top 30cm soil layer (SEW30).SEW30 is the sum of all daily values (in cm) by which watertables are closer than 30cm to the soil surface (in units of centimetre days [cm.days]). For example: 12 days rising to 10cm below the surface (20cm above the 30cm depth measured) is 12 x 20 = 240cm days.

Table 8.4: Waterlogging classes using the SEW30 index

Waterlogging classes using the SEW30 index	
Waterlogging class (in terms of drainage)	SEW30 index (cm.days in an average growing season)
Well drained	<30
Moderately well drained	30–100
Moderately drained	100–250
Imperfectly drained	250–500
Moderately poorly drained	500–1200
Poorly drained	1200–2500
Very poorly drained	>2500

Source: Soil Guide: A handbook for understanding and managing agricultural soils

Effects of waterlogging on soils

(i) **Lack of aeration:** Waterlogging removes air from the soil pores resulting in a saturated condition. In absence of air, plant roots degenerate and crops can die. Aerobic microorganisms cannot survive resulting in reduced microbiological activity which are necessary for formation of plant nutrition. Waterlogging also increases acidity build up which is harmful to most food crops.

(ii) **Reduced soil temperature:** waterlogging lowers soil temperature which restricts root development, depresses biotic activity in the soil resulting in lowered rate of production of available nitrogen, hampering seed germination and seedling growth. Reduction of soil temperatures; results in stunted growth and reduced production of nitrogen.

(iii) **Salinization:** when water from lower soil layers which may contain salts is brought up to the soil surface by capillary action, Salinity build up is increased which creates high salinization and deposits of sodium salts in the soil at or near the ground surface a.This deposition of sodium salts in the soil at or near the ground surface may be toxic or lead to the formation of alkaline conditions.

(iv) **Inhibiting activity of soil bacteria:** In waterlogged soil, when oxygen in the soil is depleted, aerobic organisms die or become dormant. Thus bacteria tend to reduce normal biotic activity and this affects root development.

(v) **Denitrification:** Denitrification occurs because of the competition for nitrogen by the soil micro-organisms that thrive in waterlogged soil and reduction in numbers of nitrifying organisms due to lack of aeration. There is reduction of nitrogen in the soil which affects plant nutrients uptake.

(vi) **Retards cultivation:** In waterlogged soil normal cultivation is difficult.

Effects of waterlogging on crops

(i) Delayed cultivation operations: Normal cultivation operations of tillage and ploughing are adversely affected due to presence of excess water in waterlogged soil.

(ii) Aquatic weeds. Water-loving wild plants grow vigorously and have competition with the crops. It suppresses the growth of useful crops.

(iii) Water logging damage to germination, seedlings and rapidly growing plants.

(iv) Diseased crops: Waterlogged conditions cause physiological disease to crops. Decay of roots, external symptoms on the foliage and fruits are common.

(v) Loss of cash crops: Cash crops desired to be grown cannot be cultivated and the land is restricted to few crops like paddy rice.

(vi) Low yields: Maturity period of crops is reduced resulting in low yields. The yield of crops is adversely affected if the water table is -

Water table	Adversely affected crops
within 90 cm	Sugarcane
60 cm	Rice
90 cm	gram and barley
90-125	Wheat
120 cm	Fodder
125 cm	maize and cotton
210-240 cm	Lucerne

(vii) Oxygen depletion: In waterlogged soil, plant roots are unable to get normal circulation of air; the level of oxygen declines and that of carbon dioxide increases which results in wilting and ultimately death of plants. The rotting of the plant roots under conditions of reduced supply of oxygen, causes yellow colour to leaves. The lack of oxygen in waterlogged soil causes precipitation of Manganese which is toxic to plants.

Management of waterlogged /Flooded soils

- **Levelling of land**: Levelling of land in many wetlands removes water by run off.
- **Drainage**: Drainage removes excess water from the root zone which is harmful for plant growth. Land can be drained by surface drainage, sub-surface drainage and drainage well methods.

- **Flood control measures:** Construction of bunds may check water flow from the rivers to the cultivable lands.
- **Plantation of tree having high transpiration rate:** In transpiration process the underground water is consumed by trees, which lowering the ground water table. Transpiration rate in certain tree likes Eucalyptus, acacia, zyzyphus is very high.
- **Raised Bed Farming:** Plant roots can't survive in water logging conditions because water logging creates anaerobic environment. But in raised bed cultivation practices an artificial aerobic condition created by cultural practices.
- **Selection of crops and their proper varieties:** Certain crops like rice, water nut, jute and sesbania can tolerate water logging upto some extent.

Table 8.5: Submergence tolerant crops and their varieties

Situation of water logging	Crop	Variety
1. Deep water	Rice	Br14, janki, Sudha, Jalmagna, Jaladhi, Madhukar
2. Shallow water	Waternut, Rice	Pankaj, Radha, Vaidehi, Rajshree
3. Flood(water-logged for short period)	Rice	BR 13, Janki
4.Water table near the surface and sometimes water-logging	Rice	Mahsuri, Jaishree, Sita, Kanak, Sujata
	Sugarcane	BO 3, 29
	Jowar	T85
	Bajara	T55
	Jute	JRC 212, JRO 632

References

Buresh, R. J., Reddy, K. R. and Van Kessel, C. 2008. Nitrogen Transformations in Submerged Soils. American Society of Agronomy, 677 S. Segoe Rd., Madison, WI 53711, USA.

Cho, D. Y. and Ponnamperuma, F. N. 1971. Influence of soil temperature on the chemical kinetics of flooded soils and the growth of rice. *Soil Sci.,* 112: 184-194.

Das, D. K. 2017. Introductory soil science (Reprint book)

IRRI, 1967. Annual Report International Rice Research Institute, Los Bafios, Manila, Philippines.

IRRI, 1969. Annual Report International Rice Research Institute, Los Bafios, Manila, Philippines.

Kirk, C. 2004. The Biogeochemistry of Submerged Soils.

Mati, B. M. What You Need to Know About Waterlogging in Agricultural Lands Training Notes

Munch, J., Hillebrand, T., Ottow, J.C.G. 1978. Transformation in the Feo/fedratio of pedogenetic iron oxides affected by iron-reducing bacteria. *Can. J. Soil Sci.* 58: 475-486.

Ponnamperuma, F. N. 1965. In The Mineral Nutrition of the Rice Plant, Johns Hopkins Press, Baltimore, Maryland, pp. 295-328.

Ponnamperuma, F. N. 1967. Soil Sci. In Dekker, New Yark, 103: 90-100.

Ponnamperuma, F. N. 1976. Temperature and chemical kinetics of flooded soils. Climate and Rice. The International Rice Research Institute, Los Banos, Philippines.

Ponnamperuma, F. N. 1979. Soil problems in the IRRI Farm. Seminar Report. The International Rice Research Institute, Los Banos, Philippines.

Ponnamperuma, F. N., Martinez, E. M., and Loy, T. A. 1966a. Soil Sci. 101: 421-431.

Ponnamperuma, F. N., Tianco, E. M., and Loy, T. A. 1967. Soil Sc. 103: 374-382.

Ponnamperuma, F. N.,Tianco, E.M, and Loy, T. A. 1966b. Soil Sci. 102: 408-413.

Ponnamperuma, F.N. 1972. The chemistry of submerged soils. *Advances in Agronomy*, 24: 29-96.

Ponnamperuma,F.N.,Yuan,W. L, and Nhung, M. T. 1965. Nature (London), 207: 1103-1104.

Sahai,V.N. Fundamentals of Soil(In book Olomu,M.O.,C.J.Racz, and C.M.Cho. 1973): Effect of flooding on the Eh,pHand concentrations of Fe and Mn in several Manitoba soils. *Soil Sci. Soc.Am. Proc.,* 37: 220-224.

Sahrawat, K. L. 2005. Fertility and organic matter in submergedrice soils. International Crops Research Institute for the Sem-Arid Tropics, India, pp. 324- 502.

Sarkar, A., Saha, M. and Saha, J. K. 2019. Management strategies for waterlogged soils in agriculture. Harit Dhara 2(1) January – June, 2019.

Soil Guide: The dynamic changes in a submerged soil with emphasis on pH, redox potential and reduction of manganese and iron, the parameters chosen for the side-effects of pesticides (review of literature). In a handbook for understanding and managing agricultural soils, Bulletin 4343.

Wickham, T.H. and Singh, V.P., 1978. Water movement through wet soils. In: Soils and rice. International Rice Research Institute, Los Banos, Philippines, pp. 337-357.

Wiley, C., Michel, V., Harm, G., Van der Geest, G., Daan van, M., Kess, W. and Herman, J. P. H. 2004. Automated and contimuous redox potential measurements in soil. *Environ. Qual.,* 33: 1562 – 1567.

9

Polluted Soils and Their Reclamation and Management

Abhay Kumar, Sunita Kumari, Ram Gopal and Vikas Rena

Introduction

Environment pollution is a burning topic of the day. Air, water and soil are being polluted alike. Soil being a "universal sink" bears the greatest burden of environmental pollution. It is getting polluted in a number of ways. There is importance in controlling the soil pollution in order to preserve the soil fertility and increase the productivity. Pollution may be defined as an undesirable change in the physical, chemical and biological characteristics of air, water and soil which affect human life, lives of other useful living plants and animals, industrial progress, living conditions and cultural resources. A pollutant is something which harmfully interferes with health, comfort and environment of the people. Most pollutants are introduced in the environment by waste, accidental discharge, sewage or else. They are by-products or residues from the production of something useful. Due to this our valuable natural resources like air, water and soil are getting polluted. The basis of agriculture is Soil. All crops for human food and animal feed depend upon it. We are losing this important natural resource by the accelerated erosion to some extent. In addition to this the huge quantities of man-made waste products, sludge and other product from new waste treatment plants, and even polluted water are also causing or leading to soil pollution. In order to preserve the fertility and the productivity of the soil.

Soil pollution probably represents the most faced problem in environmental pollution. This is because soil is a point of concentration and recovery of toxic compounds, salts, chemicals, disease causing agents, or radioactive materials, which have adverse effects on plant growth and animal health. Soil pollutants can contaminate water, water infiltration is the movement of water from the soil surface into the soil profile and soil is valuable resources that support cultures and plant life. Soil pollution is the cause of decrease in the productivity of soil due to the presence of soil pollutants. Moreover soil pollutants have an adverse effect on the physical, chemical, and biological properties of the soil and reduce its productivity.

Main causes of soil pollution are industrial activity, especially since the amount of mining and manufacturing has increased, pesticides and fertilizers, agricultural activities which are full of chemicals that are not completely degradable in nature and are widely utilized around the world; waste disposal, where there is also a large amount of industrial and municipal waste that is dumped directly into landfills without any treatment; and accidental oil spills, where oil leaks can happen during storage and transport of chemicals. Main effects of soil pollution are effect on health of humans, effect on growth of plants, decreased soil fertility and toxic dust.

Soil pollution is defined as the build-up of persistent toxic compounds, salts, chemicals, radioactive materials, or disease causing agents in soil, which have adverse en effects on plant growth and animal health.

There are many different ways that soil can become polluted, like as:

- Seepage from a landfill
- Discharge of industrial waste into the soil
- Percolation of contaminated water into the soil
- Rupture of underground storage tanks
- Excess application of pesticides, herbicides or fertilizer
- Solid waste seepage

The most common chemicals involved in causing soil pollution are

- Petroleum hydrocarbons
- Heavy metals
- Pesticides
- Solvents
- Inorganic toxic compounds-Inorganic residues in industrial waste cause serious problems regarding their disposal. They contain metals which have high potential for toxicity. Industrial activity also emits large amounts of arsenic fluorides and sulphur dioxide (SO_2). Fluorides are found in the atmosphere from superphosphate, phosphoric acid, aluminium, steel and ceramic industries. Sulphur dioxide emitted by factories and thermal plants may make soils very acidic. These metals cause leaf injury and destroy vegetation. Copper, mercury, cadmium,

lead, nickel, arsenic are the elements which can accumulate in the soil, if they get entry either through sewage, industrial waste or mine washings. Some of the fungicides containing copper and mercury also add to soil pollution. Smokes from automobiles contain lead which gets adsorbed by soil particles and is toxic to plants. The toxicity can be minimized by building up soil organic matter, adding lime to soils and keeping the soil alkaline.

- Organic wastes-Organic wastes of various types cause pollution hazards. Domestic garbage, municipal sewage and industrial wastes when left in heaps or improperly disposed seriously effect health of human beings, plants and animals. Organic wastes contain borates, phosphates, detergents in large amounts. If left untreated they will affect the vegetative growth of plants. The main organic contaminants are phenols and coal. Asbestos, combustible materials, gases like methane, carbon dioxide, hydrogen sulphide, carbon monoxide, sulphur dioxide, petrol are also contaminants. The radioactive materials like uranium, thorium, strontium etc. also cause dangerous soil pollution. Fallout of strontium mostly remains on the soil and is concentrated in the sediments. Decontamination procedures may include continuous cropping and use of chelate amendments. Other liquid wastes like sewage, sewage sludge, etc. are also important sources of soil problems.

Causes of soil pollution

- Deforestation
- Indiscriminate use of pesticides, insecticides and herbicides
- Dumping of large quantities of solid waste
- Soil erosion
- Pollution due to urbanization
- Inorganic Fertilizers
- Livestock wastes
- Inferior Irrigation Practices
- Solid Waste

Effects of soil pollution

Agricultural

- Reduced soil fertility

- Reduced nitrogen fixation
- Increased erodibility
- Larger loss of soil and nutrients
- Deposition of silt in tanks and reservoirs
- Reduced crop yield
- Imbalance in soil fauna and flora

Industrial

- Dangerous chemicals entering underground water
- Ecological imbalance
- Release of pollutant gases
- Release of radioactive rays causing health problems
- Increased salinity
- Reduced vegetation

Urban

- Closing of drains
- Inundation of areas
- Public health problems
- Pollution of drinking water sources
- Foul smell and release of gases
- Waste management problems

Environmental

- Contaminated soil is used to grow food, the land will usually produce lower yields
- Without plants on the soil will cause more erosion
- Pollutants will change the makeup of the soil and the types of microorganisms that will live in it.
- Soil pollution to change whole ecosystems

Control of soil pollution

The following steps have been suggested to control soil pollution. To help prevent soil erosion, we can limit construction in sensitive area. In general we would need less fertilizer and fewer pesticides if we could all adopt the 3R's: Reduce, Reuse, and Recycle. This would give us less solid waste.

- Extraction and separation techniques
- Thermal methods
- Chemical methods
- Microbial treatment methods
- Reducing chemical fertilizer and pesticide use- Applying bio-fertilizers and manures can reduce chemical fertilizer and pesticide use. Biological methods of pest control can also reduce the use of pesticides and thereby minimize soil pollution.
- Reusing of materials-Materials such as glass containers, plastic bags, paper, cloth etc. can be reused at domestic levels rather than being disposed, reducing solid waste pollution.
- Recycling and recovery of materials
- Aforestation
- Solid waste treatment

The use of bioremediation as a cleanup strategy for contaminated environments has increased due to its viability and cost effectiveness (Adetitun *et al.*, 2018; Akande *et al.*, 2018 and Alam *et al.*, 2018). Bioremediation uses biological agents such as fungi, bacteria, and green plants (phytoremediation) to remove, mineralize or neutralize hazardous substances in soil (Adetitun *et al.*, 2018; Akande *et al.*, 2018 and Abbaslou *et al.*, 2017). Bioremediation is divided into two types which are natural attenuation and engineered bioremediation to allow environmental conservation (Abbaslou *et al.*, 2017 and Maletic *et al.*, 2013).

The soils which possess characteristics that make them uneconomical for the cultivation of crops without adopting proper reclamation measures are known as problem soils.

Types of problem soils

Physical problem soils- contains

- Slow permeable soils/Impermeable soils and their management
- Soil surface crusting

- Sub soil hard pan
- Shallow soils
- Highly permeable soils
- Heavy clay soils
- Fluffy paddy soils

Reclamation and management physical pollute soils

- The soils should be ploughed uniformly
- Usual preparatory cultivation is carried out after compaction
- Shallow ploughing should be given and crops can be raised
- Providing open/ subsurface drainage
- Huge quantity of sand /red soil application to change the texture
- Contour/compartmental bunding to increase the infiltration
- Soil is at optimum moisture regime then ploughing is to be done.
- Application of soil conditioners like vermiculite to reduce runoff and erosion
- Incorporation of organics: Addition of organics namely FYM/ composted coir pith/press mud at 12.5 t ha^{-1} found to be optimum for the improvement of the physical properties
- Formation of ridges and furrows: For rainfed crops, ridges are formed along the slopes for providing adequate aeration to the root zone.
- Formation of broad beds: To reduce the amount of water retained in black clay soils during first 8 days of rainfall, broad beds of 3-9 m wide should be formed either along the slope or across the slope with drainage furrows in between broad beds.
- Lime or gypsum @ 2 t ha^{-1} may be uniformly spread and another ploughing given for blending of amendment with the surface soil.
- Application of clay soil up to a level 100 t ha^{-1} based on the severity of the problem and availability of clay materials
- Application of organic materials like farm yard manure, compost, press mud, sugar factory slurry, composted coir pith, sewage sludge etc
- Scraping the surface soil by tooth harrow will be useful

- Bold grained seeds may be used for sowing on the crusted soils
- More number of seeds/hill may be adopted for small seeded crops
- Sprinkling water at periodical intervals may be done wherever possible
- Growing shallow rooted crops
- Frequent renewal of soil fertility
- Resistant crops can be grown
- Growing crops that can withstand shallowness (Mango, country goose berry, fig, tamarind, ber and cashew etc)
- Crop rotation with green manure crops like Sunhemp, sesbania, daincha, kolinchi etc
- Frequent irrigation with low quantity of water
- Irrigation should be stopped 10 days before the harvest of rice crop

Chemical Problem soils

- Salt-affected soils (Saline soils, Alkali/Sodic soils, Saline-alkali soils)
- Acid soils
- Acid sulphate soils
- Calcareous soil
- Man made polluted soils

Reclamation and management polluted saline soils

- The reclamation of saline soils involves basically the removal of salts from the saline soil through the processes of leaching with water and drainage. Provision of lateral and main drainage channels of 60 cm deep and 45 cm wide and leaching of salts could reclaim the soils. Sub-surface drainage is an effective tool for lowering the water table, removal of excess salts and prevention of secondary salinization.
- Soil/cultural management
- Irrigation management
- Fertilizer management
- Crop choice/Crop management

Reclamation and management of polluted alkali/sodic soils

- Physical amelioration- It contains like Deep ploughing, Providing drainage, Sand filling, Profile inversion etc
- Chemical amelioration

- Direct Ca suppliers: Gypsum, calcium carbonate, phospho-gypsum etc.
- Indirect Ca suppliers: Elemental Sulphur, sulphuric acid, pyrites, FeSO4 etc
- Distillery spent wash- Distillery spent wash is acidic (pH 3.8-4.2) with considerable quantity of magnesium. About 2 lakh litres of distillery spent wash can be added to an acre of sodic soil in summer months.
- Distillery effluent- Distillery effluent contains macro and micronutrients. Because of its high salt content, it can be used for one time application to fallow lands, About 20 to 40 tons per ha of distillery effluent can be sprayed uniformly on the fallow land.
- Crop choice- Agroforestry systems like silviculture (Ber, tamarind, sapota, wood apple, date palm), silvipasture etc. can improve the physical and chemical properties of the soil along with additional return on long term basis. Some grasses like *Bracharia mutica* (Para grass) and *Cynodon dactylon* (Bermuda grass) etc.
- Lime requirement of an acid soil may be defined as the amount of liming material that must be added to raise the pH to prescribed value. Shoemaker *et al.* (1961) buffer method is used for the determination of lime requirement of an acid soil.
- Bioremediation- Bioremediation can be defined as any process that uses microorganisms, fungi, green plants or their enzymes to return the environment altered by contaminants to its original condition.

Biological Problem soils contains

- Soil organic compound and microbial population
- Earthworms
- Soil Enzymes

Reclamation and management of biological problem soils

- Improving Carbon Levels- No tillage farming, continuous application of manure and compost, and use of summer or winter cover crops. Burning, harvesting, or otherwise removing residues decreases soil organic compound.

- The practices that boost earthworm populations are Tillage Management (no-tillage, strip tillage, ridge tillage), Crop Rotation (with legumes) and Cover Crops, Manure and Organic by-product application, Soil reaction (pH) management and proper irrigation or drainage
- Improving Soil Respiration:The rate of soil respiration under favorable temperature and moisture conditions is generally limited by the supply of soil organic matter. Agricultural practices that increase SOM usually enhance soil respiration.
- Improving Enzyme Activity: Organic amendment applications, crop rotation, and cover crops can be used to enhance enzyme activity.

Nutritional problem soils- It contains

- Eroded soils
- Wind erosion

Best Management Practices: that are used to control erosion factors of both wind and water are

- Crop rotation
- Contour cultivation
- Strip cropping
- Terraces
- Contour terraces, and parallel terraces
- Grassed Waterways
- Diversion structures
- Drop structures
- Riparian strips
- No-tillage planting
- Strip Rotary
- Till Planting
- Annual Ridges
- Chiseling
- Disking

References

Abbaslou, H. and Bakhtiari, S. 2017. Phytoremediation potential of heavy metals by two native pasture plants (*Eucalyptus grandis* and *Ailanthus altissima*) assisted with AMF and fibrous minerals in contaminated mining regions. *Pollution*, 3(3): 471-486.

Adetitun, D., Akinmayowa, V., Atolani, O. and Olayemi, A. 2018. Biodegradation of jet fuel by three Gram negative Bacilli isolated from kerosene contaminated soil. *Pollution*, 4(2): 291–303.

Akande, F., Ogunkunle, C. and Ajayi, S. 2018. Contamination from petroleum products: Impact on soil seed banks around an oil storage facility in Ibadan, South-West Nigeria. *Pollution*, 4(3): 515-525.

Alam, A. R., Hossain, A. B. M., Hoque, S. and Chowdhury, D. A. 2018. Heavy metals in wetland soil of Greater Dhaka District, Bangladesh. *Pollution*, 4(1): 129-141.

Bech, J. 2018. Reclamation and management of polluted soils: options and case studies. *Journal of Soils and Sediments*, 18: 2131-2135.

Maletic, S., Dalmacija, B. and Srdan, R. 2013. Petroleum Hydrocarbons Biodegradability in Soil-Implications for Bioremediation. http://dx.doi.org/10.5772/50108

Ranieri, E., Bombardelli, F., Gikas, P. and Chiaia, B. 2016. Soil Pollution Prevention and Remediation. *Applied and Environmental Soil Science*, pp. 1-2.

10

Calcareous Soils and Their Reclamation and Management

Swati Shabnam, Soumitra Sankar Das and Kerobim Lakra

Introduction

Calcareous soils are those that contain high levels of calcium carbonate ($CaCO_3$) that affects soil properties related to plant growth, such as soil water relations and the availability of plant nutrients (Elgabaly, 1973), and give effervescence visibly releasing CO_2 gas when treated with dilute 0.1 N hydrochloric acid. The pH of calcareous soil is > 7.0 and also regarded as an alkaline (basic) soil.

Calcareous soils have often more than 15% $CaCO_3$ in the soil that may occur in various forms (powdery, nodules, crusts etc.). Soils with high $CaCO_3$ belong to the Calcisols and related calcic subgroups of other soils. They are relatively widespread in the drier areas of the earth. They occupy >30% of the earth's surface, and their $CaCO_3$ content varies from just detectable up to 95% (Marschner, 1995).

Formation of calcareous soils

Calcareous soils occur naturally in arid and semi-arid regions because of relatively little leaching. They also occur in humid and semi-humid zones if their parent material is rich in $CaCO_3$, such as limestone, shells or calcareous glacial tills, and the parent material is relatively young and has undergone little weathering. Some soils that develop from calcareous parent material can be calcareous throughout their profile. This will generally occur in the arid regions where precipitation is scarce. It has been estimated that these soils comprise over one-third of the world's land surface area (Taalab *et al.* 2019, Aboukila *et al.* 2018).

The soil is formed largely by the weathering of calcareous rocks and fossil shell beds like varieties of chalk, marl and lime stone and frequently a large amount of phosphates. Soils are often very fertile, thin and dry. They are found in large part of arid and semiarid regions, which may prove very fertile when sufficient moisture for crops is available.

Soils can also become calcareous through long term irrigation with water contains small amounts of dissolved $CaCO_3$ that can accumulate with time. Calcareous soils can contain from 3% to >25% $CaCO_3$ by weight with pH values with a range of 7.6 to 8.3.

Problems of cultivating calcareous soils

Cultivation of calcareous soils presents many challenges, such as low water holding capacity, high infiltration rate, poor structure, low organic matter (OM) and clay content, low CEC, loss of nutrients via leaching or deep percolation, surface crusting and cracking, high pH and loss of nitrogen (N) fertilizers, low availability of nutrients particularly phosphorous (P) and micronutrients, and a nutritional imbalance between elements such as potassium (K), magnesium (Mg) and calcium (Ca) (El-Hady and Abo-Sedera, 2006, FAO, 2016). Under such severe conditions, desired yield levels are difficult to attain.

Calcareous soil lacks the OM necessary for optimum soil function as cropland. The addition of organic amendments improves soil chemical and physical properties, initiates nutrient cycling, and provides a functioning environment for vegetation.

The potential productivity of calcareous soils is high where adequate water and nutrients can be supplied. The high calcium saturation tends to keep the calcareous soils in well aggregated form and good physical condition. However, where soils contain an impermeable hard pan (petricalcic horizon) they should be deeply ploughed in order to break the pan. This should be followed by the establishment of an efficient drainage system. Furrow irrigation is better than basin irrigation on slaking calcareous soils. On undulating lands, contour and sprinkler irrigations are better options than flood irrigation. Drip irrigation may also be practiced. Calcareous soils generally have low organic matter content and lack nitrogen. Nitrogen fertilizer may be applied any time from just before planting up to the time the plant is well established. Application of nitrogen through side-dressing to the growing crop is an efficient way of nitrogen application. Care should be exercised so as not to apply nitrogen close to the seed as it may prevent germination. Ammoniac sources of nitrogen and urea should not be left on the surface of calcareous soils, since considerable loss of ammonia through volatilization may occur, and they should be incorporated in the soil instead.

Phosphorous is often lacking in calcareous soils. Amounts to apply depend on how deficient the soil is and the crop requirements. Excess applied phosphorus may lead to deficiency of zinc or iron. To be effective on calcareous soils, applied phosphorus fertilizer should be in water soluble form. Band application of phosphate is more effective as compared to broadcast application. Application at the time of seeding has been found to be most appropriate since phosphorus is required mostly during the younger stages of plant growth.

Calcareous soils usually suffer from a lack of micronutrients, especially zinc and iron. Zinc deficiency is most pronounced in maize, especially under high yield intensive cultivation systems. Zinc sulphate is an effective zinc source and is the most popular form in use. For soil application, zinc sulphate is broadcast and incorporated in soil. A single application lasts for several years. Foliar applications of zinc are used on fruit trees. Heavy applications of animal manure are helpful in preventing deficiency of iron and zinc.

Major problems related to calcareous soils can be listed as-

Crust formation

Crust formed on the soil surface is one of the most important components of the soil. This is true in the case of arid, semiarid and the moderate environment as well (Rahmonov, 2006). Gillette proved with his experiments carried out in wind tunnel that surface covered with crust is less vulnerable for wind erosion (Gillette *et al.*, 1982). There are two main types of surface crusts: one of them is formed out in natural way and the other one is formed out by anthropogenic effects. The former one can be divided into two other groups: the one that is formed by moisture (this is called as physical) and the one that has a biological origin (Goossens, 2004). The latter one forms out as a result of the cementing effect of different algae, lichens and bacteria living in soil. These micro-plants can bind soil particles and don't let them start to move and protect them from the bombarding effect of other grains. Excretion of organic matter- for example polysaccharides- can increase the strength of such type of crust.

The biological crusts are important in the arid and semiarid regions, because the vegetation cover is not developed enough to provide the stability of surface (Eldridge, 1998). Investigations carried out in desert regions established that the biological crust has more strength and provide better protection than the other one evolved by the effect of precipitation (Rajot *et al.*, 2003; Zhang *et al.*, 2006).

The formation of crusts is a problem in the carbonate-rich soils newly put under cultivation. Crusting which takes place at the soil surface hinders seedling rate of emergence and percentage. The adverse effect of crust depends on their strength and thickness. Hillel (1986) found that emergence of bean seedlings in fine sandy loam soil was reduced from 100 to 0% when crust strength increased from 108 to 273 mbar whereas the emergence of sweet corn was prevented only when crust strength exceeded 1200 mbar. Lemos and Lutz (1957) showed that increase in the silt or soil fractions less than 0.1mm, the

2:1 type clay mineral, compaction by rainfall and soil puddling increase crust strength. They showed that the calcareous soils they used became severely crusted and compacted after repeated flooding. They stated that the action of water disperses the soil and reorients the soil particles. Thus, a close compact arrangement resulted when the soil was dried.

Ammonia volatilization

In the soil solution, the ammonia gas is in equilibrium with ammonium ions according to the following reversible reaction:

$$NH_4^+ + OH^- \rightarrow H_2O + NH_3 \uparrow$$

From the above equation it is obvious that ammonia volatilization will be more pronounced at high pH levels (pH>7.5). On the other hand, ammonia-gas-producing amendments will drive the reaction to the left, raising the pH of the solution in which they are dissolved. When a moist soil dries, the water is moved from the right-hand side of the equation, again driving the equation to the right, encouraging ammonia volatilization. Compared with NH_3 volatilization loss from N fertilizers, NH_3 loss from N mineralized from organic N is relatively small. Urea hydrolysis can increase soil pH thus encourage NH_3 volatilization. Therefore, in neutral and acidic soils, NH_4^+-N containing fertilizers are less subject to volatilization loss than urea-containing fertilizers. However, in calcareous soils, soil solution pH is buffered at about 7.5, a condition favor volatilization loss for even NH_4^+ - containing fertilizers, for example:

$$(NH_4)_2SO_4 + CaCO_3 \rightarrow 2NH_3 + H_2O + CO_2 + CaSO_4$$

Since $CaSO_4$ is only slightly soluble, the reaction proceeds to the right and NH_3 volatilization is favored. Similar reactions occur with other NH_4^+ -containing fertilizers that produce insoluble Ca precipitates (e.g., $[NH_4]_2HPO_4$). In comparison, volatilization losses are not as great if the NH_4^+ -containing fertilizers (e.g., NH_4NO_3, NH_4Cl) produce soluble Ca reaction products. Ammonia volatilization losses also increase with increasing fertilizer rate and with liquid compared with dry N sources. Volatilization of NH_3 is much greater with broadcast applications compared with subsurface or surface band methods. Immediate incorporation of broadcast N greatly reduces the NH_3 volatilization potential.

Soil colloids, both clay and humus, adsorb ammonia gas, so ammonia losses are greatest where low quantities of these colloids are present or where the ammonia is not in close contact with the soil. For these reasons, ammonia loss can be quite large from sandy soils and from alkaline or calcareous soils, especially when the ammonia-producing materials are left at or near the soil surface and when the soil is drying out. High temperatures, as often occur on the surface of the soil, also favor the volatilization of ammonia. Incorporation of manure and fertilizers into the top few centimeters of soil can reduce ammonia losses by 25-75% from those that occur when the materials are left on the soil surface.

By the reverse of the ammonia loss mechanism described above, both soil and plants can absorb ammonia from the atmosphere. Thus, the soil-plant system can help cleanse ammonia from the air, while deriving usable N for plants and soil microbes. Forests may receive a significant proportion of their N requirements as ammonia carried by wind from fertilized cropland or cattle feedlots located many kilometers away.

Precipitation of soluble phosphate

Phosphorus (P) is an essential macronutrient, being required by plants in relatively large quantities (~0.2 to 0.8%). Potassium and nitrogen are the only mineral nutrients required in larger quantities than P. Providing adequate P to plants can be difficult, especially in calcareous soil. Alkaline soil is defined as soil with pH greater than neutral, typically 7.5 to 8.5. Calcareous soil is defined as having the presence of significant quantities of free excess lime (calcium or magnesium carbonate). Lime dissolves in neutral to acid pH soil, but does not readily dissolve in alkaline soil and, instead, serves as a sink for surface adsorbed calcium phosphate precipitation.

The bioavailability of P is strongly tied to soil pH. The formation of iron and aluminum phosphate minerals results in the reduced solubility of P in strongly acidic soil, improving as pH approaches nearly neutral. This maximum solubility and plant availability of P at pH 6.5 declines again as the pH increases into the alkaline range. This effect of reduced P availability in alkaline soil is driven by the reaction of P with calcium, with the lowest solubility of these calcium phosphate minerals at about pH 8. The presence of lime in alkaline soil further exacerbates the P availability problem. The lime in calcareous soil reacts with soil solution P to form a strong calcium phosphate bond at the surface of the lime. These calcareous soils are common in arid and semi-arid regions with little rainfall.

The resulting effect of low P solubility in calcareous soil is relatively poor fertilizer P efficiency. Plants grown in these conditions can be stunted with shortened internodes and poor root systems due to P deficiency. Deficiency symptoms are sometimes observed as a darkening of the leaf tissue, although it is more common to observe yield loss with no readily seen symptom.

Simply adding fertilizer P at "normal" rates and with conventional methods may not result in optimal yield and crop quality. Several fertilizer P management strategies have been found to improve P nutrition for plants grown in alkaline and calcareous soil, namely: 1) relatively high P fertilizer rates, 2) concentrated P fertilizer bands, 3) complexed P fertilizer, 4) slow release fertilizer P, 5) cation complexing P fertilizer, and 6) balancing P with other nutrients. These methods may be used alone or in various combinations to effectively supply P to plants growing in calcareous soil.

Adsorption and precipitation of P in calcareous soils take place; it is difficult to distinguish between the two mechanisms. The solid phase of adsorbed P in the soils is very complex (Wandruszka, 2006). Due to the surface adsorption and precipitation process, all the fertilizer P applied to soil is not available to plants. The experiments have shown that application of soluble P and concentration of labile P decreased rapidly with time due to reaction with soil, either due to precipitation or adsorption process by Fe oxides and hydroxides in acidic soils and $CaCO_3$ in calcareous soils (Pant and Warman, 2000).

Heredia *et al*., (1997) reported that extractable values were not correlated with adsorption measures. Adsorption and extractions were related to different soil particles that varied according to great group under consideration. Association with different particles was also variable for different great group. Furthermore, coarse texture soil was found to release more P than fine textured soil (Pothuluri *et al*., 1991). Cumulative release of P had a highly positive relationship with supply parameters irrespective of P level. Soil with slightly lower pH, clay content and CEC did not differ appreciably in respect of adsorption, desorption and supply parameter of P, owing to their similar mineralogical make up. Phosphate reported to be adsorbed on the $CaCO_3$ surfaces and amorphous or crystalline calcium phosphate precipitate. In addition, P in calcareous soils may be adsorbed on the surface of other soil constituent like hydrous oxide, clay and organic matter. The relative importance of these forms of phosphate is contributing to labile P has not been completely defined. Nevertheless, these P forms must provide for the regeneration of solution phase P. Generally, the amount of labile P depends greatly on the amount and the adsorption energy of the P on the soil surfaces and secondly, on the nature and solubility of the dominant compounds occurring in the soil. Although there is no direct proof that $NaHCO_3$ extracts all the exchangeable P first and then from non-exchangeable specific surface area, P adsorption was decreased and desorption increased, but varied for a given soil with soil P status.

Management of calcareous soil

The calcareous soil can be managed in the following ways:

(i) Tillage Operation: Light (sandy) calcareous soil develops large number of pore spaces due to flocculation. These type of soils have poor water-holding capacity. Therefore, such type of soil needs compaction by plank or roller to increase the water- holding capacity.

(ii) Application of Organic Manure: When sufficient amount of farm yard manure, compost and green manure are added, the amount of carbon dioxide and acid increases and as a result pH of soil decreases.

(iii) Use of Chemical Fertilizer: Fertilizer management in calcareous soils is different from that of non calcareous soils because of the effect of soil pH on soil nutrient availability and chemical reactions that affect the loss or fixation of some nutrients. The presence of CaCO3 directly or indirectly affects the chemistry and availability of nitrogen (N) Phosphorus (P), Magnesium (Mg), Potassium (K), Manganese (Mn), Zinc (Zn) and iron (Fe). The availability of copper (Cu) also is affected.

Application of acid forming fertilizers such as ammonium sulphate and urea fertilizers, sulphur compounds, organic manures and green manures is considered as effective measures to reduce the pH of soil to neutral pH value.

Availability of phosphorus is low in calcareous soil. To increase the availability of P, the phosphatic fertilizers should be used in the following manner:

(a) Phosphatic fertilizers should be used near the roots of plant.

(b) Use of phosphatic fertilizers in ball form also increases its availability.

(c) May be used in split dose.

(iv) Use of Micronutrients: Addition of micronutrients like, zinc, copper, iron would be helpful in increasing the yield.

References

Aboukila, E. F., Nassar, I. N., Rashad, M., Hafez, M. and Norton, J. B. 2018. Reclamation of calcareous soil and improvement of squash growth using brewer's spent grain and compost. *Journal of the Saudi Society of Agricultural Sciences* 17: 390-397.

Eldridge, D.J. 1998. Trampling of microphytic crusts on calcareous soils and its impact on erosion under rain-impacted flow. *Catena*, 33: 221-239.

Elgabaly, M.M. 1973. Reclamation and management of the calcareous soils of Egypt. In: FAO Soils Bulletin 21, Calcareous soils: report of the FAO/UNDP Regional Seminar on Reclamation and Management of Calcareous Soils, Cairo, Egypt, 27 Nov - 2 Dec 1972. FAO Soils Bulletin No. 21, pp. 123–127.

El-Hady, O. A. and Abo- Sedera, S. A. 2006. Conditioning effect of composts and acrylamide hydrogels on a sandy calcareous soil. II-Physioco-bio-chemical properties of the soil. *International Journal of Agricultural Biology,* 8(6): 876-884.

FAO. 2016. FAO Soils Portal: Management of Calcareous Soils.

Gillette, D. A., Adams, J., Muhs, D. and Kihl, R. 1982. Threshold friction velocities and rupture moduli for crusted desert soils for the input of soil particles into the air. *Journal of Geophysical Research*, 87: 9003-9015

Goossens, D., 1998. Effect of soil crusting on the emission and transport of wind-eroded sediment: field measurements on loamy sandy soil. *Geomorphology,* 58: 145-160.

Heredia, O., Giuffre, L. and Rotondaro, R. 1997. Adsorbed phosphorus: Relation with Bray and Olsen P in Great Groups of soils in Buenos Aires Province. Revista-de-la-Facultad-de-Agronomia-Universidad-de-Buenos-Aires. 17(3): 253 – 262.

Hillel, P., 1986. Biochemistry of Fern Spore Germination: Protease Activity in Ostrich Fern Spores. *Plant Physiol.*, 80: 992-996.

Lemos and Lutz 1957. Soil crusting and some factors affecting it. *Soil Sci. Am. Proc.,* 21: 485-491.

Marschner, H. 1995. Mineral Nutrition of Higher Plants. Academic Press, London.

Pant, H., and Warman, P. 2000. Phosphorus release from soils upon exposure to ultra-violet light. *Comm. Soil Sci. and Plant Anal.,* 31(3-4): 321-329.

Pothuluri, J., Whitney, D. and Kissel, D. 1991. Residual value of fertilizer phosphorus in selected Kansas's soils. *Soil Sci. Soc. Am. J.,* 55: 399-404.

Rahmonov, O., Czylok, A. and Simanauskiene, R. 2006. The significance of biological soil crust in regeneration ecosystems of anthropogenic bare sand. Anthropogenic aspects of landscape transformations 4. pp. 88-95.

Rajot, J.L., Alfaro, S.C. and Gaudichet, A. 2003. Soil crusting on sandy soils and its influence on wind erosion. *Catena* 53: 1-16.

Taalab, A.S., Ageeb, G.W., Hanan, S. S. and Safaa, A. M. 2019. Some Characteristics of Calcareous soils. *A review Middle East Journal of Agriculture Research*, 8(1): 96-105.

Wandruszka, R., 2006. Phosphorus retention in calcareous soils and the effect of organic matter on its mobility, http://www.geochemicaltransactions.conv'content /7/1/6.

Zhang, Y.M., Wang, H.L., Wang, X.Q., Yang, W.K. and Zhang, D.Y. 2006. The microstructure of microbiotic crust and its influence on wind erosion for a sandy soil surface in the Gurbantunggut Desert of Northwestern China. *Geoderma* 132: 441-449.

11

Desert Soils and Their Reclamation and Management

Shikha Verma and Reshma Shinde

Introduction

Desert soils are the soils of arid regions that remain dry for most part of the year and support sparse xerophytes. The amount of water received by precipitation is far less than what is lost by evapotranspiration. Most deserts receive an average precipitation of less than 400 mm. Because of extreme imbalance between evapotranspiration and precipitation, many desert soils contain salts.

Distribution

Deserts occupy about one-third of the areas of Africa and Australia, 11% of Asia and about 8% of the Americas (Boul *et al.* 1997). Desert soils include both cold and hot desert. In India, desert soils occur in three agro-ecological regions (AER). The cold arid desert of extreme north (Leh, Ladakh region of Jammu & Kashmir), the hot arid desert of north western India (Rajasthan, southern parts of Haryana and Punjab and northern parts of Gujarat) and the tropical arid desert of the Deccan plateau in South India represents AER 1, 2 and 3, respectively. The three different desert regions in India have characteristic climatic and soil conditions.

1. Cold arid eco-region with shallow skeletal soils

This covers an area of 15.2 million hectares (mha) in the north-western Himalayas (Ladakh and Gilgit districts). The mean annual temperature is less than 8 °C while the mean annual rainfall is less than 150 mm. This ecoregion has aridic and crycic soil moisture and soil temperature regime, respectively. The soils are shallow, sandy, calcareous, alkaline in reaction and low to medium in organic matter content.

2. Hot arid eco-region with desert and saline soils

This AER covers the western parts of Rajasthan, south-western parts of Haryana and Punjab, the Kachchh peninsula and northern parts of Kathiawar peninsula (Gujarat) accounting for 31.9 m ha of the geographical region of India. The region has hot summers and cool winters (arid), with the mean annual precipitation being less than 400 mm. The ecosystem is characterized by aridic soil moisture and hyperthermic soil temperature regimes. Soils are sandy in texture, moderately calcareous in nature and alkaline in reaction. The scanty rainfall in the region means heavy water deficit, with acute droughtiness at the time of grain formation. The salinity of the soil results in physiological droughts that occur frequently. There is nutrient imbalance of nitrogen, phosphorous, zinc and iron in particular.

3. Hot arid eco-region with red and black soils

Characterized by hot and dry summers and mild winters, this eco-region covers an area of 4.9 m ha in the region of Deccan plateau in the South India. The rainfall is erratic and ranges from 400-500 mm against annual potential evapotranspiration of 1800-1900 mm. Severe drought conditions persist throughout the year with a gross water debit of1500-1600 mm every year. The region has aridicustic soil moisture and isohyperthermic soil temperature regimes. Shallow and medium red soils are dominant and these are slightly acidic and non-calcareous. Deep, clayey black soils are slightly alkaline and calcareous.

Classification

Typical desert soils have been classified in the order **Aridisol** (sub ordes- cryids: Aridisol in cold areas; salids: accumulation of salts; Gypsids: accumulation of gypsum; calcids: accumulation of carbonates; druids- accumulation of silica and cambids: translocation and/or transformation of materials) and others in **Entisol** (Suborders- Psamments: deep sandy entisols with low water holding capacity and orthents: entisol on recent erosional surface that have shallow sand cover).

The classification of such soils occurring in different temperature regimes is given (Table 11.1).

Table 11.1: Classification of desert soils in India

Region	Soil Regimes		Classification
	Moisture	Temperature	
Cold, arid	Hyper-aridic	Frigid	Cryothents, Cryids, (with inclusions of Cryepts in valleys
Hot, arid	Typic/Hyper-aridic	Hyperthermic/ Megathermic	Calcids, Gypsids, Cambids(Torripsamments)
Tropical,arid	Typic/Hyper-aridic	Iso-mega-hyperthermic	Torrerts

Source: Sehgal, 2014

General properties of desert soils

The desert soils are basically sandy in texture with low clay content (5 to 10 %). They are mostly single grained structure with poor aggregation of particles. Quartz and plagioclase minerals are commonly found in desert soils. They are generally brown, light brown, yellow, gray or reddish in colour depending on factors like climate, minerals present, moisture and oxidation and hydration state of iron compounds. Red soils are common in hot deserts while cold deserts show gray colour. Physical weathering predominates over chemical weathering in the development of desert soils.

Desert soils are variable in nature depending on parent material and climate .They may be saline, sodic or calcareous in nature. Most desert soils are saline due to the fact that limited rainfall is not sufficient to leach the soluble products of weathering to deeper layers and hence the salts accumulate in the soil. Further during dry weather salts move up with the water and tend to concentrate on the surface as the water evaporates. Soil sodicity/alkalinity is due to sodium salts in soil solution or in the exchange complex or both. The presence of Na-clay and natric horizon makes the soils susceptible to dispersion. They are moderately to strong alkaline in reaction with pH range from 7.9-9.0. Values of pH 8.5 or greater almost invariably indicate an exchangeable sodium percentage of 15 or more and the presence of alkaline earth carbonates (Richards, 1954). Carbonate accumulation is considered to be an important pedogenic feature of desert soils (Boul *et al.,*1997)

The soils are low in organic matter content and poor in nutrient and water holding capacity. They are basically poor in nitrogen status. The cation exchange capacity of desert soils low in clay and organic matter is usually low. The high soil pH values lead to the precipitation of P, Fe, Zn, Cu and Mn, mostly in forms that are not available for plant uptake. Occurrence of surface crusts and desert pavements are also the features of desert soils.

Constraints of desert soils in relation to crop production

- Low water and nutrient holding capacity due to sandy texture.
- Poor fertility status
- Physical (wind erosion) and chemical (salinization, alkalization) degradation.
- Low organic matter content due to sparse vegetation and rapid oxidation because of high temperature.
- Salt toxicity in saline desert soils. High concentration of salts reduces the ability of plants to absorb water which interferes with their growth.
- P-fixation in calcareous desert soils.
- Micronutrient deficiency in high pH desert soils.
- Problem of dispersion and low permeability in case of alkaline desert soils where Na ion predominates.
- Presence of surface crusts act as a barrier in seedling emergence and inhibits the infiltration of water.

Reclamation and Management of Desert soils

Water Management: Since, potential evapotranspiration greatly exceeds precipitation in desert regions, moisture conservation, efficient utilization of precipitation and stored soil moisture play a key role for judicious management of desert soils.

Rainwater management: Rainfall in desert areas occurs often under the form of heavy rainstorm and is erratic in nature, followed by intense surface runoff that lasts only few hours or days. Efficient rainwater management in such areas needs in situ and ex situ rainwater harvesting.

In situ rainwater harvesting: It is based on changing soil and water management techniques with the aim to improve infiltration, water holding capacity and soil fertility. The runoff in the sloping fields can be captured in the field by employing various types of barrier and can be stored in soil for immediate use by the crops. In-situ rainwater harvesting is generally carried out through various agronomic and engineering measures where the rainfall is collected where it falls. Some of the widely adopted techniques are bunding, terracing, contour farming, deep ploughing , mulching, pitting etc.

Ex situ rainwater harvesting: This involves diverting the runoff and storing it in a natural or artificial reservoir for future use. This can be achieved through dugout ponds, diversion bunds, tanks, *etc*. Diversion of perennial surface/

subsurface water source is practiced through check-dams, nala-bunding, percolation tanks etc.

Improved designs like underground cintern called tanka, improved nadi lined with low density polyethylene sheets, silt trap and khadins have been developed and standardized for Jaisalmer with less than 200 mm rainfall (Narain, 2008). Khadin (earthen bunds for rainwater harvesting in cropped fields) is a traditional method of rainwater harvesting in Jaisalmer against an embankment where cropping is practical on receding moisture level.

Control of seepage and evaporation loss from storage reservoirs: Seepage and evaporation losses from storage reservoir in deserts are of great concern due to light textured sandy soils, high temperature, low relative humidity and high wind speeds. Seepage loss can be checked by providing linings at the bottom and side surfaces of the reservoirs. A mixture of bentonite clay and soil is used for lining the surface of the pond, which seals the soil pores. A layer of masonry lining of brick and cement is also a common practice.

Evaporation losses from the reservoirs can be checked by decreasing the surface area by increasing the depth of the tank. Shading of water storage tanks and addition of chemicals which forms protective layer on surface to reduce evaporation is also practiced.

Control of evaporation from soil surface: Moisture loss through evaporation from the soil surface can be minimized by i) controlling the energy supply to the evaporation site which can be achieved by modifying albedo through change of colour or structure of soil surface or by shading the surface; ii) by decreasing the conductivity or diffusivity of the profile particularly of surface zone (via tillage and mulching practices).

Covering or mulching the surface with vapor barriers or with reflective materials can reduce the intensity of radiation and wind on the soil surface. Common mulch materials used for the purpose are sawdust, crop residues, straw, leaves, other litter etc. Specially designed papers and plastic, impermeable plastic films and asphalt prays are also used as mulch with considerable success in desert areas.

At the later phase of drying process i.e. in relatively dry soil, reduction of water loss by evaporation depends on decreasing the rate of diffusivity or hydraulic conductivity of the soil profile. Deep tillage by way of increasing the range of variation of diffusivity with changing water content may reduce the rate at which the soil can transmit water towards the surface and ultimately reduce the loss of water from the deeper layers of the soil profile by evaporation.

Irrigation management: Efficient and economic utilization of stored rainwater in desert regions calls for suitable irrigation techniques. Some of the ways for better irrigation management in desert soils are:

- A desert soil generally suffers from the problem of salinity. Hence care must be taken to use good quality salt free water for irrigation.
- Frequent and light irrigation in sandy desert soils minimizes the seepage loss of water.
- Scheduling of irrigation should be done at moderate soil temperature because irrigation on hot soil may cause evaporation loss of water. So in hot summer months irrigation is advisable during late afternoon or at night and in cold winter month during late morning or early noon.
- Suitable method of irrigation must be adopted for increased water use efficiency. Flood irrigation is not feasible for desert soils as most of the water will either evaporate or go deep in the soil rather than being available for the plant. Suitable methods of irrigation in desert soils are sprinkler or drip irrigation, but both have some limitations. Sprinkler irrigation is suitable for short statured crops, shallow sandy soils and for saline soils to leach salts, but it is not suitable during high wind velocity. Drip irrigation is well suited to areas of acute water shortage as it minimizes the losses of water by deep percolation, evaporation and runoff. The limitation with drip method is that it is not suitable for saline soils because the salts cannot be leached with this method.
- Since, sprinkler and drip irrigation methods are costly, modified irrigation technique for irrigating sandy soils is by porous pitcher method. Porous pitcher filled with water are inserted in a pit or near the base of the plant and covered with lid. Water that comes out of the pitcher irrigates the root zone of the crop.
- Acidifying irrigation water has shown to be beneficial and has given good production results in calcareous arid soils (Bashour, 2009).

Soil conservation

Management of soil erosion by wind

Wind erosion is very common in desert regions due to strong wind regime, low atmospheric humidity, high solar radiation and single grain structure of sandy soils. Besides, its common occurrence in hot arid desert of north western India (Rajasthan, southern parts of Haryana and Punjab and northern parts of Gujarat), wind erosion is also prevalent in the coastal areas where sandy soils predominate as well as in the cold desert regions of Leh. The destructive effect of wind erosion includes loss of soil fertility, crop damage and pollution. Basic principles of wind erosion control include reduction in wind velocity near the ground surface, increasing the resistance of soil surface and trapping of saltating soil particles. For controlling wind erosion following have been found effective:

- Strip cropping of grass with crop
- Creation of rough, cloddy surface
- Creation of permanent grass stips across the prevailing wind direction
- Stubble mulch tillage in which crop residues are left on the surface to reduce wind velocity and trap eroding soil
- Pebble and gravel mulches can be used to control wind erosion
- Avoiding the cultivation of soil while it is dry
- Windbreaks/ Shelterbelts (living windbreaks viz. *Cassia siamea*, *Acacia tortilis* etc.) as barrier for protection from winds.

Sandune stabilization

The formation of dunes may be prevented or slowed down by adopting measures like:

(a) Reducing the velocity of wind by erecting barriers (physical stabilization)

(b) Restricting sand movement through chemical sprays (chemical stabilization)

(c) Encouraging the development of vegetation that binds sand (biological stabilization).

Stabilization of sand dunes with vegetation may be accompanied by establishing grasses and then reforesting. Grasses having strong extensive root system should be used for the purpose. Once stabilized the dunes can be planted with shrubs and trees.

Crops and cropping systems

Crops with deep root system, draught resistant and salt tolerant crops and species are well suited to desert soils. Examples of such crops include maize, millets, pulses, barley, sorghum, cotton etc. Improved varieties of pearl millet (CZP-9802, CZP-IC-923) , moth bean (Maru moth, CAZRI-Moth-1, CAZRI-Moth-2, CAZRI-Moth-3) , cluster bean (Maru-Guar) and horsegram (Maru-Kulthi-1) are recommended for desert soils. Low water requiring short duration varieties is advisable in desert soils to maximize production under small quantity of water. CAZRI has screened a number of strains of desert grasses, viz., *Cenchrus ciliaris* (Anjan), *Lasiurus sindicus* (Sewan), *Cenchrus setigerus* (Dhaman) , *Dichanthium annulatatum* and *Panicum antidote* (Narain, 2008). Legume based intercropping with legumes like mothbean, clusterbean etc. have been found not only to improve the yield but increase the organic matter in desert soil (Aggarwal and Praveen Kumar, 1993).

Soil fertility management

The approach has the following components:

i. **Addition of organic matter:** Addition of organic matter through different sources like FYM, crop residues etc improves the soil moisture storage capacity, stabilizes the soil structure, resist erosion and supply essential nutrients to the crop.

ii. **Efficient fertilizer use:** Desert soils are particularly deficient in nitrogen. Hence, the crop production in such soils require external application of splitting of nitrogen dose into 2 or 3 splits and use of slow release nitrogenous fertilizers can help in minimizing the N losses . The volatilization losses of nitrogen applied as urea can be sufficiently reduced by the placement of fertilizer deep (15cm) in the soil (Aggarwal and and Kaul, 1978). Further work at CAZRI, Jodhpur has revealed that losses of nitrogen from urea in volatilization can be substantially reduced with the mixing of sulphur with urea (Aggarwal *et al.* 1987). In calcareous desert soils, phosphatic fertilizers should be placed in narrow bands to minimize P- fixation. Application and incorporation of potassium before or at planting is better than side dressing in calcareous desert soils. This is because the side dressed K is less likely to move to the root zone because of its low mobility in calcareous soils. Deficiencies of iron, zinc, copper, manganese are associated with high pH desert soils and their deficiencies can be ameliorated with foliar sprays.

iii. **Improved Cropping system:** The incorporation of fodder and grain legumes in the intermixed cropping systems of cereals, grasses, and oilseeds has both direct and indirect effects on soil fertility. Introduction of legumes in cropping system as intercrops enhances the fertility of soil by means of nitrogen fixation and binds the soil particles through their extensive root system thereby improving the soil structure and thus protecting against erosive action of the wind.

Reclamation and management of specific conditions

Desert soils are highly variable in nature depending on parent material and climate, Soils derived from sandstone become sandy in texture, and those derived from limestone become calcareous (Bashour, 2009). Salinization is common in desert soils due to low precipitation which is insufficient to leach the salts to deeper layers as well as high evaporation rate which tends to concentrate the salts in the soil. Further, continuous addition of sodium containing salts via fertilizers or irrigation water tends to saturate the exchange sites with Na resulting in the formation of sodic soils. Each of these soil types requires special management.

Management of salinity

- Leaching with good quality water to remove the salts from the root zone of soils.
- Mechanical means to remove the salts that have accumulated on the soil surface in order to improve permeability and drainage within the soil profile viz. scraping, sub soiling , profile inversion and deep ploughing.
- Use of salt free irrigation water.
- Maintaining the soil moisture level near field capacity to dilute the salt concentration in soil solution.
- Use of organic manures.
- Control of evaporation from soil surface.
- Growing salt tolerant crops (barley, sugarbeet, wheat, sorghum, maize etc).

Management of sodicity/alkalinity

- Application of amendments like gypsum, iron pyrites, sulphur etc.
- Cultivation of alkali tolerant crops.
- Addition of molasses.
- Addition of organic matter.

Management of calcareous soils

Desert soils containing sufficient calcium carbonate are calcareous in nature. Such soils can be managed in following ways:

- Compaction by means of roller or plank to decrease the pore spaces in order to increase the water holding capacity.
- Addition of organic manures. Decompostion of organic manures releases sufficient carbon dioxide and organic acids as a result of which pH of soil decreases.

- Band placement of phosphatic fertilizers to minimize its fixation.
- Addition of micronutrient (Fe, Zn, Cu) containing fertilizers as these are deficient in high pH soils.

Conclusions

Desert soils are inherently fertile as they contain high amount of weatherable minerals, but because of aridity, chemical weathering is limited to release weathering products (Sehgal,2014). Lack of sufficient precipitation is one of the major constraints encountered in desert areas. With efficient soil and water management practices, it is possible that the desert soils can be profitably used for growing crops.

References

Aggarwal, R.K. and Kumar, P. 1993. Sustainable productivity through organic manures in arid land ecosystem. Proc. Of International symposium on Environmental Degradation, CAZRI, Jodhpur.

Aggarwal, R. K. and Kaul, P. 1978. Loss of Nitrogen as Ammonia Volatilization from Urea on Loamy Sand Soil of Jodhpur. *Ann. Arid zone,* 17(2): 242-45.

Aggarwal, R. K. Raina, P. and Kumar, P. 1987. Ammonia Volatilisation Losses from Urea and their Possible Management for Increasing N Use Efficiency in Arid Region. *Jr. Arid. Env.,* 13(2): 166-60.

Boul, S. W., Hole, F. D., McCracken, R. J. and Southard R. J. 1997. Soil Genesis and Classification. Panima Publishing Corporation, New Delhi. pp. 259-272.

Isam, I. B. 2009. Desert Reclamation and management of drylands: Fertility aspects. In: Encyclopedia of land use, land cover and soil sciences (eds. Willy H. Verheye.), Eolss Publishers Co. Ltd., Oxford, United Kingdom. 5:116-136.

Narain, P. 2008. Dryland management in arid ecosystem. *Journal of the Indian Society of Soil Science,* 56(4): 337-347.

Richards, L. A. 1954. Diagnosis and Improvement of Saline and Alkali Soils. Govt. Printing Office, Washington D. C.

Sehgal, J. 2014. A Text Book of Pedology: Concepts and Applications. Kalyani Publishers, New Delhi, pp.150-260.

12

Black Cotton Soils and Their Reclamation and Management

S. Firdous and Parthasarathi T.

Introduction

Black cotton soils are formed by the solidification of molten magma; it covers a large area of the Deccan plateau. The depth of this soil is very high in the lower part of the Krishna – Godavari basin. The clay fraction is very high which is responsible for the heavy waterlogging in the area of heavy rainfall and become more vulnerable to soil erosion. The fine fraction of soil not allows draining out through the soil profile and at the same time the rate of evaporation is low. During summer soils suffer from moisture stress and it becomes very hard but at the same time during the rainy season moisture content very high. this imposes severe constraints on conventional agriculture use as a result most of the land remains fallow and subjected to soil erosion due to heavy rain. The nutrient status of the soil is quite good in Vertisol but these constraints limit crop production. To manage black cotton soil for crop production it is advisable to use animal-powered devices for surface soil drainage, Counter cultivation, adoption of a new cropping system will be helpful for better crop production.

The distribution of black cotton soil in India is mainly found in the central and western region. These are inorganic clays characterized as high shrinkage and swelling properties with medium to high compressibility and plasticity. These soils become very plastic when moist and very hard upon drying and therefore, very difficult to cultivate and manage. On drying, there is vertical movement of soil forms cracks of varying depth. 2:1 type of clay mineral (smectitic group) is common in this soil having more than 30% clay content and at least 50 cm soil depth. It has high CEC, low organic matter, calcium, potassium, and magnesium content is high but nitrogen content is relatively low.

In India black soil is known as *"regur"*, *"tirs"* in Morocco *"margalitic"* soils in Indonesia, and *"black turf"* in Africa. These soils are also comparable with Grumosols and Prairie of USA and Chernozems of Russia but there is significant differences in their physico-chemical properties.

Distribution of black cotton soil

Black soil falls under the Vertisols soil order, and it has been found all over the world. It covers about 2% i.e. 257 Mha of the total ice-free earth surface and 73 million ha of the total geographical area in India *i.e.* 28 million ha which are true Vertisols. The distribution of most of these soil (80%) are located in the states of Maharashtra, Gujarat, Madhya Pradesh, and Andhra Pradesh, 13% found in karnataka and Tamil Nadu, and the rest in other adjoining states (Kanwar *et al.,* 1952). The Black soils are formed by the solidification of lava stretch over the Deccan plateau and in the lower part of Krishna and Godavari Basin black soils are quite deep. In the current review from available datasets, there is an expansion of the area to about 117 M ha (Mandal *et al*., 2018).

The rainfed farming system is common in this area because of the short rainy season and seasonally dry climatic belt. Based on the rainfall, Virmani and his co-worker (1981) have categorized Vertisol (Soil order) of India into two climatic regions, i) areas with relatively dependable rainfall where the mean annual rainfall ranges from 750 to 1250 mm or more and ii) areas with relatively undependable and low rainfall, ie. <750 mm mean annual rainfall. The dependable rainfall zones are considered by marked reliability of rainfall incidence at short intervals, whereas zones with unpredictable rainfall show the recurrent occurrence of long drought periods.

Pedogenic development

Black cotton soils are found in semiarid to sub-humid climates with distinct annual wet and dry spells. The formation of this soil by the chemical weathering of basic igneous rocks, for example granite, basalt, andesites, norite, schists, gneisses which comprises calcium-rich feldspars and dark minerals and sedimentary materials such as shales, limestones, slates, etc. It weathered from the clay-rich parent materials predominantly smectite type consisting of high coefficient of expansion and contraction. Clemente *et al*., (1996) and Murthy *et al*., (1982) reported that most vertisols are originated from the Cenozoic era, and its development occurs in different topographical situations such as from the summit to the valley bottom of the terrain.

There is a lack of horizon development in Vertisols because they have a "vertic horizon" which is a clayey subsurface horizon with slickensides or edge shape structure aggregates. These are characterized by argillipedoturbation (disruption and mixing) caused by wetting and drying, expansion, and contraction (swell- shrink) of the soil mass (Johnson, 1987).

A steady churning occurs in the pedon due to expansion or swelling of soil and it creates huge pressures within the soil. Wedge-shaped aggregates i.e. a typical soil structure formed in the surface soil and blocks-like structure form the subsoil due to vertical mixing (argillipedoturbation). In the subsoil, one block slips over another block resulting in formation of a unique polished surface known as slickensides. Swelling and shrinkage also causes the development of micro-topographic structures known as gilgai, that is formed by mass movements within the soil and throwing of the underlying material towards the surface. The upper horizon is commonly dark grey to black whitish and the lower horizon colour varies from grey-brown, reddish-brown to a whitish colour.

Characteristics of black cotton soils

- The clay content in black soil is high i.e. 30-80%. It consists of smectitic group of clay minerals which has a high coefficient of swelling and shrinkage properties.
- The cation exchange capacity (CEC range is 30-60 cmol (p+) kg^{-1} soil) and is high base saturation.
- It has high water and nutrient holding capacity but a large fraction of these nutrients are not available to plant because of the presence of high clay content which holds the nutrient firmly.
- Its colour is very dark due to the presence of titaniferous magnetite mineral and clay humus complex.
- It has low permeability and poor drainage property due to the occurance of high smectitic clay minerals.
- Black soil have high moisture retention capacity and it becomes very sticky and plastic in wet condition and very difficult to plough whereas under the dry condition it becomes very hard. Under this situation, soil get shrinks and create a wide and deep crack.
- Silica: sesquioxide ratio of the clay is 3.0 -3.5.
- There is no distinct horizon due to the churning / mixing of soil.
- The pH of the black soil ranges between 7.8-8.7 but 9.5 also noticed under sodic condition.

Constraints

The main constraints in black soils are limited workable soil moisture range, low infiltration rate and poor drainage condition which leads to waterlogging and high erosion and runoff soil loss during a heavy downpour and during summer experience moisture stress-causing drought. These soils have low organic carbon, nitrogen, sulphur, phosphorus and sometimes Zn also found deficient. Water holding capacity is very high because of the high amount of clay content and after irrigation, it becomes more susceptible to salinity and sodicity in the subsoil.

Management

Rainfed agriculture mainly depends on rain for water sources. Infiltration rate is very poor in black soil causing waterlogging and erosion problems during the rainy season. It is a serious concern and to solve this problem, natural topography and drainage system is a good option.

It includes:-

1. Development of small watershed, a natural framework to conserve soil moisture and reduce erosion.
2. Contour cultivation is also helpful for minimizing runoff and erosion by offering more time to penetrate rainwater through innumerable miniature barriers.
3. Vertical mulching is also helpful for enhancing the crop yield.
4. Addition of moderate amount of nitrogen and phosphatic fertilizers after that dry sowing of 2:1 ratio of sorghum/maize –pigeon pea intercropping about one week before the onset of the monsoon and adoption of disease-resistant crop varieties advisible.
5. Intercropping is also helpful to increase the yield like 2:1 ratio of sorghum/pigeon pea or pearl millet/pigeon pea and 1:1 sorghum/mung bean intercropping.
6. Advance planting of rabi crop by 20-30 days is helpful for enhancing crop yield and maintaining the availability of soil moisture and nutrients.
7. Crop sequencing also gives a good result in yield enhancement such as mung bean/ Sorghum/maize - saffloweror sorghum/ maize –chickpea.

References

Coulombe, C.E., Wilding, L.P. and Dixon, J.B. 1996. Overview of vertisols: characteristics and impacts on society. In: Sparks, D.L. (Ed.), Advances in Agronomy, vol. 57. Academic Press, New York, pp. 289–375.

Johnson, D.L. and Watson-Stegner, D. 1987. Evolution model of pedogenesis. *Soil Science* 143, 349–366.

Kanwar, J.S., Kampen, J. and Virmani, S.M. 1982. Management of Vertisols for Maximizing Crop Production-ICRISAT Experience. International Crops Research Institute for the Semi-Arid Tropic (ICRISAT), Patancheru, Andhra Pradesh, India.

Mandal, C., Mandal, D.K., Prasad, J., Sarkar, D., Chandran, P., Tiwary, P., Patil, N.G., Obi Reddy G.P., Lokhande, M.A., Wadhai, K.N., Dongare, V.T., Sidhu, G.S., Sahoo, A.K., Nair, K.M., Singh, R.S., Pal, D.K., Ray, S.K. and Bhattacharyya, T. 2018. Revision of Black Soil Map of India for Sustainable Crop Production. In: Conference on National Seminar on Geospatial Solutions for Resource Conservation and Management. Karnataka State Remote Sensing Application Centre, Bangalore.

Murthy, R.S., Bhattacharjee, J.C., Landey, R.J. and Pofali, R.M. 1982. Distribution, characteristics and classification of vertisols.Vertisols and Rice Soils of the Tropics, Symposia Paper II, 12th International Congress of Soil Science, New Delhi. Indian Society of Soil Science, pp. 3–22.

Sehgal, J. L. 2008. A textbook of Pedology concepts and applications. New Delhi: Kalyani Publishers, pp. 274-278

13

Irrigation Water Their Quality and Standards

Swati Shabnam

Introduction

Irrigated agriculture is dependent on an adequate water supply of usable quality. The water quality used for irrigation is essential for the yield of crops, maintenance of soil productivity, and protection of the environment as well. The quality of irrigation water is defined with respect to its effect on plant growth, soil properties, soil biological equilibrium, and irrigation technology. Conceptually, water quality refers to the characteristics of a water supply that will influence its suitability for a specific use, i.e. how well the quality meets the needs of the user. Specific uses have different quality needs and one water supply is considered more acceptable (of better quality) if it produces better results or causes fewer problems than an alternative water supply. For example, good quality river water which can be used successfully for irrigation may be unacceptable for municipal use without treatment to remove its sediment load. Similarly, snowmelt water of excellent quality for municipal use may be too corrosive for industrial use without treatment to reduce its corrosion potential.

Suitability of water for irrigation

The suitability of water for long term use in irrigation is determined by its quality. All water contains salts, some which are harmful (sodium, bicarbonate) and some which improve soil structure (calcium, magnesium). Over time, soils take on the chemical properties of the water used on them. In irrigation water evaluation, emphasis is given on the chemical and physical characteristics of the water, and rarely are any other factors considered important.

The suitability of irrigation water depends upon several factors, such as, water quality, soil type, plant characteristics, irrigation method, drainage, climate and the local conditions. The integrated effect of these factors on the suitability of irrigation water (SI) can be qualitatively expressed by the relationship:

SI = QSPCD, where

Q = quality of irrigation water, that is, total salt concentration, relative proportion of cations, etc;

S = soil type, texture, structure, permeability, fertility, calcium carbonate content, type of clay minerals and initial level of salinity and alkalinity before irrigation;

P = salt tolerance characteristics of the crop and its varieties to be grown, and growth stage;

C = climate, that is, total rainfall, its distribution and evaporation characteristics; and

D = drainage conditions, depth of water table, nature of soil profile, presence of hard pan or lime concentration and management practices.

These factors act interactively. For example; in a particular climate, all the factors enumerated above are likely to vary and interact either positively or negatively in relation to salt accumulation and degree of harmful effect on soil properties and crop growth. As such, a single suitable criterion is hard to be adopted for widely varying conditions. However, a general broad guideline has been developed for use by the field practitioners.

Besides these factors, presence of some ions in water such as calcium, sulphate, potassium, and nitrate is favorable for crop growth, as even saline water can be used in presence of these ions.

Quality of irrigation water

The term "water quality" includes the individual and combined effects of the substances present in the water. The possibility of using a given water supply for irrigation purposes is determined to a large extent by its constituents. There are two types of constituents in any water:

1. **Suspended particles-** these are mainly silt and clay particles which imparts turbidity and colour to water used. Other than this there may be organic material suspended in water.
2. **Dissolved matter-** there are salts and other solids dissolved in water which gives water its chemical properties.

Water Quality Indices

The water quality indices or chemical properties which can be evaluated in lab are listed below:

pH

The pH is the concentration of hydrogen ions (H^+) and hydroxyl ions (OH^-) present in the water. It is used to determine the acidic, basic or neutral behavior of water. The pH values ranges from 1 to 14. If pH of water is less than 7 then it is called acidic water whereas, pH equal to 7 as neutral and more than 7 is called the basic nature water. pH highly affects the efficiency of coagulation and flocculation process (Kahlown *et al.,* 2006).

Electrical Conductivity (EC)

Electrical conductivity is the measure of salts in water. Salts are composed of positively and negatively charged ions, which dissociate when salt is dissolved in water. These dissolved ions are also called as electrolytes. Thus, the electrical conductivity (EC) of water is defined as the capacity of water to transmit the electric current. It depends on the dissolved ions in the water and their charge and movement and is directly related to the concentration of ions in the water. These conductive ions come from dissolved salts and inorganic materials such as alkalis, chlorides, sulfides and carbonate compounds.

When the EC of water is high, it shows that there is high concentration of ions in the water. The EC indicates the number of total solids in water and is dependent on the temperature of water. The measurement of EC at 25°C temperature is considered as reference. The electrical conductivity of water also affects the plant growth.

Conductivity Units : Conductivity is usually measured in micro- or milli siemens per centimeter (uS/cm or mS/cm). It can also be measured in micro mhos or milli mhos/centimeter (umhos/cm or mmhos/cm), though these units are less common. The basic unit of EC in SI units is Siemens m^{-1} (previously mhos m^{-1}). Formerly water salinities were expressed in micro mhos cm^{-1}. Now mS m^{-1} is the standard expression for EC of both soils and waters.

The conversion is:

1 mS m^{-1} = 0.01 m mho cm^{-1} =10 μmho cm^{-1}

Frequently EC is multiplied by a factor to obtain total soluble salts (mass/volume) as an expression of salinity. There is however no unique factor that can be applied and the factor will vary with composition and concentration (George 1983).

Total Dissolved Solids (TDS)

The salinity behavior of water is indicated by total dissolved solids (TDS). TDS is the measure of the amount of material dissolved in water including carbonate, chloride, bicarbonate, phosphate, sulfate, nitrate, sodium, calcium, magnesium, organic ions etc. Water contains anions (negatively change ions) and cations (+ve changes ions). The relationship between total dissolved solids and EC is:

TDS (mg/L) = EC (dS/m) × K (1)

Where

K = 640 in most cases (for EC: 0.5 -5 dS/m) or

K = 735 for mixed waters or

K = 800 for EC > 5 dS/m

The above relationship in most cases is applied for EC ranging from 0.5 to 5 dS/m (Kahlown and Khan, 2002). The density of the water, due to high TDS concentrations, determines the flow of water into and out of an organism's cells, and hence can be harmful. Moreover, the high concentrations of TDS may also reduce water clarity, contribute to a decrease in photosynthesis, combine with toxic compounds and heavy metals, and lead to an increase in water temperature.

Total Suspended Solids (TSS)

Total suspended solids (TSS) are the fine particles consisting of microorganisms, algae, mineral particles and organic matter suspended in water. Total suspended solid is, thus, an indicator of erosion and sediment transport and load, and it absorbs heat energy from sun resulting in increase in water temperature and consequently decreases in the level of dissolved oxygen, as amount of oxygen dissolved in water is inversely proportional to the temperature of water.

Total Solids (TS)

Total solids (TS) are the combination of total suspended solids and total dissolved solids in the water and are measured in milligrams per liter (mg/L). The dissolved solids can pass through a filter of around 2-microns in size. Suspended solids are bigger in size than dissolved solids and those would not pass through a 2-micron filter (APHA, 1992). In water, the suspended and dissolved solids came from different sources such as soil erosion, sewage, fertilizer, industrial discharges, and road runoff.

Turbidity

The cloudiness in the water is known as turbidity, which is caused by dissolved or total suspended solids, and most of the time is invisible to the naked eye as smoke in air. Turbidity is the values of light absorbing or light scattering property of water. It is an important parameter as a measure of water quality. Turbidity can be caused by i.e. silt, sand and mud; bacteria and germs; as well as chemical precipitates. The turbidity is measured in Nephelometric Turbidity Units (NTU), defined by US Environmental Monitoring Standard unit.

Color

The color of water is also an important indicator to define pollutant level in water. Water color represents the type of solid material present in it. The apparent blue color of water bodies is due to selective absorption and scattering of light spectrum. Transparent water with low level of dissolved solids has blue color while yellow or brown color is due to the dissolved organic matter. Some algae produce reddish or deep yellow waters. Similarly, the water rich in phytoplankton and other algae appears as green. True color could be measured by filtering the water after removing all suspended material (CWT, 2004).

Calcium and Magnesium (Ca, Mg)

The calcium and magnesium in water comes from the decomposition of calcium and magnesium alumino-silicates, and from dissolution of limestone, magnesium limestone, magnesite, gypsum and other minerals. Calcium is an essential element for living organisms, mainly in cell physiology and mineralization of bones and shells. In water, calcium is in higher quantity than that of magnesium. Moreover, in groundwater and surface water also, the concentration of calcium by weight is very high. The total concentration of calcium and magnesium is referred as water hardness. Hard water forms precipitates on boiling. The calcium carbonate is the most dominant factor because of calcium and carbonate and referred as total hardness, measured in mg/L.

Carbonates and Bicarbonates (CO_3^{2-}, HCO_3^-)

Carbonate and bicarbonate ions are present in water and come from dissolving of carbon dioxide (CO_2) by naturally circulating waters. The atmospheric carbon dioxide is partly intercepted by photosynthesizing vegetation, which is converted to cellulose starch and related carbohydrates. The concentration of carbonates in natural waters is a function of dissolved carbon dioxide, temperature, pH, cations and other dissolved salts. Carbonate is a salt of carbonic acid, which originates from dissolving of carbonate minerals. A carbonate salt is formed when a positively charged ion, attaches to the negatively charged oxygen atoms of the carbonate ion. The bicarbonate ion (hydrogenated-carbonate ion) is an anion with a negative charge and is the conjugate acid of carbonate. The weathering of rocks contributes to bicarbonate content in water as mostly these are soluble in water and their concentration in water depends on pH of water. It is a principal alkaline constituent in almost all water sources, therefore, influences hardness and alkalinity of water. Many types of bicarbonate are soluble in water at standard temperature and pressure, particularly sodium bicarbonate and magnesium bicarbonate; both of these substances contribute to total dissolved salts which are a common parameter for assessing water quality. The concentration of carbonates and bicarbonates should be under the limit in water and if the level of carbonates and bicarbonates is increased, it may be harmful for humans, animals and plants (Arshad and Shakoor 2017).

Sodium Adsorption Ratio (SAR)

Sodium adsorption ratio (SAR) is a property that gives information on the comparative concentrations of sodium, in respect to calcium and magnesium in the water. SAR is a measure of the tendency of the irrigation water to cause the replacement of calcium (Ca) ions attached to the soil clay minerals with sodium ions (Na). Sodium clays have poor structure and develop permeability problems. The SAR can be calculated as:

$$SAR = \frac{Na}{\sqrt{0.5\ (Ca + Mg)}}$$

Where, all concentrations are in me/L. Concentrations of these ions in water samples are typically provided in milligrams per liter (mg/L). To convert Na, Ca, and Mg from mg/L to meq/L, concentration should be divided by 22.9, 20, and 12.15 respectively.

Residual Sodium Carbonates (RSC)

It is used to predict the additional sodium hazard associated with CaCO3 precipitation involve calculation of the residual sodium carbonate. RSC is another alternative measure of the sodium content in relation with calcium and magnesium. This can be calculated as:

RSC = (CO3

2- + HCO3 -) – (Ca2+ + Mg+2) (3)

Where, all concentration is in meq/L.

Problems Related to Irrigation Water Quality

Water quality can be measured in laboratory for so many parameters. But as far as growth of crops is concerned, following criteria for quality of irrigation water is most important:

a) the total concentration of soluble salts, or salinity hazard

b) the relative proportion of sodium to the other cations, Sodicity or sodium hazard

c) the bicarbonate concentration as related to the concentration of calcium and magnesium

d) the concentrations of specific elements and compounds

Salinity hazard

In most irrigated situations, the primary water quality concern is salinity levels, since salts can affect both the soil structure and crop yield. Water used for irrigation can very much vary in quality depending upon the type and quantity of dissolved salts. Irrigation water is usually drawn from surface or groundwater sources and typically contains salts in the range of 200 to 2000 ppm. These salts are carried with the water to wherever it is used. In the case of irrigation, the salts are applied with the water and remain behind in the soil as water evaporates or is used by the crop.

Many salinity problems are associated with or strongly influenced by a shallow water table (within 2 metres of the surface). Salts accumulate in this water table and become an additional source of salt that moves upward into the crop root zone frequently. Control of the existing shallow water table is thus very essential to control salinity and practice successful long-term irrigated agriculture. Water with high salinity requires large amount of water for leaching, which adds greatly to drainage problem and makes long-term irrigated agriculture nearly impossible to achieve without adequate drainage. However, if proper drainage is available, salinity control becomes simply good management to ensure that the crop is adequately supplied with water at all times and that enough water for leaching is applied to control salts within the tolerance of the crop.

The suitability of water for irrigation is determined not only by the total amount of salt present but also by the kind of salt. Various problems related to soil and cropping evolve as the total salt content increases, and special management practices may be required to maintain acceptable crop yields. The primary effect of high EC water on crop productivity is the inability of the plant to compete with ions in the soil solution for water (physiological drought). The higher the EC, even though the soil may appear to be wet, less is the water available to the plants. Significant yield reductions are observed when the salts accumulate in the root zone to such an extent that the crop is no longer able to extract sufficient water from the salty soil solution, resulting in a water stress for a significant period of time. If water uptake is greatly reduced, the plant growth rate slows down. The plant symptoms are similar in appearance to those of drought, such as wilting, or a darker, bluish-green colour and sometimes thicker, waxier leaves. Symptoms vary with the growth stage, being more noticeable if the salts affect the plant during the early stages of growth. In some cases, mild salt effects may go entirely unnoticed because of a uniform reduction in growth across an entire field. Actual yield reductions from irrigating with high EC water varies substantially with factors influencing yield reductions including soil type, drainage, salt type, irrigation system and management.

Salts that contribute to a salinity problem are water soluble and readily transported by water. A portion of the salts that accumulate from prior irrigations can be leached below the rooting depth if more irrigation water infiltrates the soil than is used by the crop during the crop season. Leaching is the key to controlling a water quality-related salinity problem. Over a period of time, salt removal by leaching must equal or exceed the salt additions from the applied water to prevent salt building up to a damaging concentration. The amount of leaching water required is dependent upon the irrigation water quality and the salinity tolerance of the crop grown.

Salt content of the root zone varies with depth. It is approximately that of the irrigation water near the soil surface while much higher at the bottom of the rooting depth. Each subsequent irrigation pushes (leaches) the salts deeper into the root zone, where they continue to accumulate until leached.

As a basic definition, salinity is the total concentration of all dissolved salts in water. These salts, or electrolytes, form ionic particles on dissolving in water, each with a positive and negative charge. While salinity can be measured by a complete chemical analysis, this method is difficult and time consuming. For example, seawater cannot simply be evaporated to a dry salt mass measurement as chlorides are lost during the process.

More often, salinity is not measured directly, but is instead derived from the conductivity measurement, as salinity is a strong contributor to conductivity. This is known as practical salinity. These derivations compare the specific conductance of the sample water to a standard such as seawater. Salinity measurements based on conductivity values are unitless, but are often followed by the notation of practical salinity units (psu).

Sodium hazard

Although plant growth is primarily limited by the salinity (EC) level of the irrigation water, the application of water with a sodium imbalance can further reduce yield under certain soil texture conditions. Irrigation with sodic water is of special concern due to sodium's effects on the soil and poses a sodium hazard. Sodic water is water with a high concentration of sodium, relative to the concentration of calcium and magnesium. Sodic water is not the same as saline water.

A high sodium ion in irrigation water affects the hydraulic conductivity (permeability) of soil and creates water infiltration problems. This is because when sodium present in the soil in exchangeable form replaces calcium and magnesium, adsorbed on the soil clays and causes dispersion of soil particles (i.e. if calcium and magnesium are the predominant cations adsorbed on the soil exchange complex, the soil tends to be easily cultivated and has a permeable and granular structure) leading to sealing of soil pores and a reduction in permeability to water flow.. Due high value of SAR, the soil becomes hard and compact when dry and resultantly, reduces the infiltration rates of water and air into the soil affecting its structure. This typically only happens to soil with a relatively high percentage of smectite clay, which is a group of clay minerals that includes montmorillonite and bentonite. This type of mineral tends to swell when exposed to water. Once a clay-dominated soil disperses, the soil will either become anaerobic (lacking oxygen), saline, or compacted/consolidated. The tendency for sodium to increase its proportion on the cation exchange sites at the expense of other types of cations (primarily calcium and magnesium) is estimated by the sodium adsorption ratio (SAR), which is the ratio of sodium concentration to the concentration of the square root of the average calcium plus magnesium concentration in either irrigation water or the soil solution (Miller and Gardiner, 2007).

This condition, termed "sodicity," results from excessive soil accumulation of sodium. Sodicity causes a decrease in the downward movement of water into and through the soil, and actively growing plants roots may not get adequate water, despite pooling of water on the soil surface after irrigation.

Many factors including salinity rates, soil texture, organic matter, cropping system, irrigation system and management affect how sodium in irrigation water affects soils. For example, sandy soils may not get damage as easily as other heavier soils when it is irrigated with a high SAR water. Similarly, the potential soil infiltration and permeability problems created from applications of irrigation water with high "sodicity" cannot be adequately assessed on the basis of the SAR alone. This is because the swelling potential of low salinity (EC) water is greater than high EC waters at the same sodium content. Therefore, a more accurate evaluation of the infiltration/permeability hazard requires using the electrical conductivity (EC) together with the SAR, as given in Table 13.1.

Table 13.1: Assessment of sodium hazard of irrigation water based on SAR and EC_w (modified from Ayers and Westcot, 1994)

Irrigation water SAR	Potential for water infiltration problems at EC_w	
	Unlikely	Likely
0-3	> 0.7	< 0.2
3-6	> 1.2	< 0.4
6-12	> 1.9	< 0.5
12-20	> 2.9	< 1.0
20-40	> 5.0	< 3.0

However, for irrigation water with high bicarbonate (HCO3) content, an "adjusted" SAR (SARADJ) can be calculated. In this case, the amount of calcium is adjusted for the water's alkalinity, is recommended in place of the standard SAR (see pH and Alkalinity section below). Your laboratory may calculate an adjusted SAR in situations where the HCO3 is greater than 200 mg/L or pH is greater than 8.5. The presence of or introduction of bicarbonate and carbonate ions in the irrigation water increases the permeability hazard as quantified by SAR. Irrigation of calcium-rich or magnesium rich soil with water containing carbonate or bicarbonate ions will form insoluble calcium and magnesium carbonate (limestone, dolomite), thereby reducing the concentration of calcium calcium and magnesium applied to the SAR calculation. This consideration in the calculation of SAR results in the adjusted SAR (SAR adj) being greater than the SAR, thereby providing a truer index of the sodicity of the water and the risk of dispersion. Most SAR adj values of irrigation waters are about 10 to 15 percent greater than the unadjusted SAR. Additionally, irrigation water with a low salt concentration and a high SAR will contribute to reduced permeability of dispersive soils eventually. It is important to know if you are dealing with SAR or SAR adjusted when interpreting results.

Toxic elements

A number of other substances may be found in irrigation water and can cause toxic reactions in plants. After sodium, chloride and boron are of most concern. In certain areas of Texas, boron concentrations are excessively high and render water unsuitable for irrigations. Boron can also accumulate in the soil. Crops grown on soils having an imbalance of calcium and magnesium may also exhibit toxic symptoms. Sulfate salts affect sensitive crops by limiting the uptake of calcium and increasing the adsorption of sodium and potassium, resulting in a disturbance in the cationic balance within the plant. The bicarbonate ion in soil solution harms the mineral nutrition of the plant through its effects on the uptake and metabolism of nutrients. High concentrations of potassium may introduce a magnesium deficiency and iron chlorosis. An imbalance of magnesium and potassium may be toxic, but the effects of both can be reduced by high calcium levels.

As water take nutrients when missed with water and on the same time when crop take water containing ionic constituents the toxicity problems raised. The high concentration of toxic elements can reduce crop growth and resulted in low crop production. The primary ionic constituents are boron, sodium and chloride. It is not necessary that the high concentration cause the damage even small quantity respond the same. The water infiltration and salinity problems can also be due to toxicity. Without transpiration, the plant is unable to live as this is the basic need of plant. In this transpiration process, the dissolved ions in water move and accumulated to the leaves. These ions block the stomata in the leaves and reduce the transpiration process which adversely affects the plant growth. It is also to note that the direct application of adsorbed toxic ions from overhead sprinkler can create toxicity problems. Toxicity from both sodium and chloride or from any one is dangers to sensitive crop such as citrus. Excessive amounts of chlorine, sodium, boron, lithium and other elements may be toxic to some crops.

Plant response to poor quality irrigation water

The amount of water transpired through a crop is directly related to yield; therefore, irrigation water with high EC reduces yield potential. Generally forage crops are the most resistant to salinity, followed by field crops, vegetable crops, and fruit crops which are generally the most sensitive. Salinity tolerances for irrigation water in different crops are given in Table 13.2.

Table 13.2: Salinity tolerances* in irrigation water for different crops (Adapted from Ayers and Westcot, 1994)

Field crops	Yield potential, EC_{iw}			
	100%	90%	75%	50%
Barley 5.0	5.0	6.7	8.7	12.0
Bean (field)	0.7	1.0	1.5	2.4
Broad bean	1.1	1.8	2.0	4.5
Corn	1.1	1.7	2.5	3.9
Cotton	5.1	6.4	8.4	12.0
Cowpea	0.9	1.3	2.1	3.2
Flax	1.1	1.7	2.5	3.9
Groundnut	2.1	2.4	2.7	3.3
Rice (paddy)	2.0	2.6	3.4	4.8
Safflower	3.5	4.1	5.0	6.6
Sesbania	1.5	2.5	3.9	6.3
Sorghum	2.7	3.4	4.8	7.2
Soybean	3.3	3.7	4.2	5.0
Sugar beet	4.7	5.8	7.5	10.0
Wheat	4.0	4.9	6.4	8.7

**Based on the electrical conductivity of the irrigation water (ECiw) measured in mmhos/cm.*

Boron is a major concern in some areas. While a necessary nutrient, high boron levels cause plant toxicity, and concentrations should not exceed a certain limit, given in Table 13.3. The tolerance of crops to sodium as measured by the exchangeable sodium percentage (ESP) is given in Table 13.4. Similarly, chloride tolerance of certain agricultural crops is given in Table 13.5.

Table 13.3: Boron sensitivity of selected plants

B concentration, mg/ L				
Sensitive		Moderately sensitive	Moderately tolerant	Tolerant
(0.5-0.75)	(0.76-1.0)	(1.1-2.0)	(2.1-4.0)	(4.1-6.0)
Peach	Wheat	Carrot	Lettuce	Alfalfa
Onion	Barley	Potato	Cabbage	Sugar beet
	Sunflower	Cucumber	Corn	Tomato
	Bean		Oats	

Source: Mass (1987)

Table 13.4: Relative tolerance of crops to ESP

Tolerant (ESP-55)	Semi-tolerant (ESP-35)	Sensitive (ESP-10)
Bermuda grass	Wheat	Cowpea
Paragrass	Barley	Gram
Rice	Oats	Goundnut
Sugarbeet	Senji	Lentil
	Berseem	Blackgram
	Sugarcane	Greengram
	Millets	Peas
	Cotton	Maize
		Cotton (at germination)

Table 13.5 Chloride tolerance of agricultural crops (Adapted from Tanji. 1990)

Crop	Maximum Cl^- concentration without loss in yield (ppm)
Bean	350
Corn	525
Flax	525
Potato	525
Sugarcane	525
Alfalfa	700
Sesbania	700
Rice	1050
Cotton	1625
Cowpea	1750
Wheat	2100
Sorghum	2450
Barley	2800

Classification of irrigation water

Several different measurements are used to classify the suitability of water for irrigation, including EC_{iw}, the total dissolved solids, and SAR. Some permissible limits for classes of irrigation water on the basis of EC values are given in Table 13.6. In Table 13.7, the sodium hazard of water is ranked from low to very high based on SAR values. Chloride classification of irrigation water is given in Table 13.8.

Table 13.6: Permissible limits for classes of irrigation water on the basis of EC values

Classes of water		Electrical conductivity μmhos*
Class 1	Excellent	250
Class 2	Good	250-750
Class 3	Permissible[1]	750-2000
Class 4	Doubtful[2]	2000-3000
Class 5	Unsuitable[2]	3000

*Micromhos/cm at 25 degrees C.
[1]Leaching needed if used
[2]Good drainage needed and sensitive plants will have difficulty obtaining stands

Source: Fipps

Table 13.7: The sodium hazard of water based on SAR values

SAR value	Sodium hazard	Remarks
1-10	Low	Use on sodium sensitive crops must be cautioned.
10-18	Medium	Amendments (such as Gypsum) and leaching needed.
18-26	High	Generally unsuitable for continuous use.
>26	Very high	Generally unsuitable for use.

Source: Fipps

Table 13.8: Classification of irrigation water based on chloride

Chloride (ppm)	Effect on Crops
Below 70	Generally safe for all plants.
70-140	Sensitive plants show injury.
141-350	Moderately tolerant plants show injury.
Above 350	Can cause severe problems

Source: Bauder et al.

Table 13.9: Classification of irrigation water based on boron (ppm)

Class of water	Crop group		
	Sensitive	Semi-tolerant	Tolerant
Excellent	<0.33	<0.67	<1.00
Good	0.33 to 0.67	0.67 to 1.33	1.00 to 2.00
Permissible	0.67 to 1.00	1.33 to 2.00	2.00 to 3.00
Doubtful	1.00 to 1.25	2.00 to 2.50	3.00 to 3.75
Unsuitable	>1.25	>2.50	>3.75

Source: Fipps

Management of poor quality irrigation water

Techniques for controlling salinity that require relatively minor changes are more frequent irrigations, selection of more salt-tolerant crops, additional leaching, pre-plant irrigation, bed forming and seed placement. Alternatives that require significant changes in management are changing the irrigation method, altering the water supply, land-leveling, modifying the soil profile, and installing subsurface drainage.

Leaching

Leaching is useful to reduce the soil salinity and its problems. The salinity level in the soil can be determined with water having no salts in it, especially the sodium. The soil of good structure and proper drainage is demanded for efficient leaching. For a soil having salinity, it is needed to determine the amount of salts free water to reclaim it. Some time, in soil, having high salinity level needs various application of fresh water to leach the salts below root zone with proper drainage. In the areas of shallow groundwater level, the leaching is impracticable; in such situation, the artificial drainage is recommended. The artificial drainage has some advantages to remove salinity from soil even with low quality water.

Soluble salts that accumulate in soils must be leached below the crop root zone to maintain productivity. Leaching is the basic management tool for controlling salinity. Water is applied in excess of the total amount used by the crop and lost to evaporation. The strategy is to keep the salts in solution and flush them below the root zone. The amount of water needed is referred to as the leaching requirement or the leaching fraction.

Excess water may be applied with each irrigation to provide the water needed for leaching. However, the time interval between leachings does not appear to be critical provided that crop tolerances are not exceeded. Hence, leaching can be accomplished with each irrigation, every few irrigations, once yearly, or even longer depending on the severity of the salinity problem and salt tolerance of the crop. An occasional or annual leaching event where water is ponded on the surface is an easy and effective method for controlling soil salinity. In some areas, normal rainfall provides adequate leaching.

Sub-surface drainage

Very saline, shallow water tables occur in many areas of Texas. Shallow water tables complicate salinity management since water may actually move upward into the root zone, carrying with it dissolved salts. Water is then extracted by crops and evaporation, leaving behind the salts. Shallow water tables also contribute to the salinity problem by restricting the downward leaching of salts through the soil profile. Installation of a subsurface drainage system is about the only solution available for this situation. The original clay tiles have been replaced by plastic tubing. Modern drainage tubes are covered by a "sock" made of fabric to prevent clogging of the small openings in the plastic tubing.

Seed placement

Obtaining a satisfactory stand is often a problem when furrow irrigating with saline water. Growers sometimes compensate for poor germination by planting two or three times as much seed as normally would be required. However, planting procedures can be adjusted to lower the salinity in the soil around the germinating seeds. Good salinity control is often achieved with a combination of suitable practices, bed shapes and irrigation water management. In furrow-irrigated soils, planting seeds in the center of a single-row, raised bed places the seeds exactly where salts are expected to concentrate. This situation can be avoided using "salt ridges." With a double-row raised planting bed, the seeds are placed near the shoulders and away from the area of greatest salt accumulation. Alternate-furrow irrigation may help in some cases. If alternate furrows are irrigated, salts often can be moved beyond the single seed row to the non-irrigated side of the planting bed. Salts will still accumulate, but accumulation at the center of the bed will be reduced. With either single- or double-row plantings, increasing the depth of the water in the furrow can improve germination in saline soils. Another practice is to use sloping beds, with the seeds planted on the sloping side just above the water line. Seed and plant placement is also important with the use of drip irrigation.

Residue management

The common saying "salt loves bare soils" refers to the fact that exposed soils have higher evaporation rates than those covered by residues. Residues left on the soil surface reduce evaporation. Thus, less salts will accumulate and rainfall will be more effective in providing for leaching.

Chemical amendments

In sodic soils (or sodium affected soils), sodium ions have become attached to and adsorbed onto the soil particles. This causes a breakdown in soil structure and results in soil sealing or "cementing," making it difficult for water to infiltrate. Chemical amendments are used in order to help facilitate the displacement of these sodium ions. Amendments are composed of sulphur in its elemental form or related compounds such as sulfuric acid and gypsum. Gypsum also contains calcium which is an important element in correcting these conditions. Some chemical amendments render the natural calcium in the soil more soluble. As a result, calcium replaces the adsorbed sodium which helps restore the infiltration capacity of the soil. Polymers are also beginning to be used for treating sodic soils.

It is important to note that use of amendments does not eliminate the need for leaching. Excess water must still be applied to leach out the displaced sodium. Chemical amendments are only effective on sodium-affected soils. Amendments are ineffective for saline soil conditions and often will increase the existing salinity problem.

References

APHA, 1992. Standard methods for the examination of water and wastewater. 18th edition. American Public Health Association, Washington, DC.

Arshad, M. and Shakoor, A. 2017. Irrigation Water Quality. *Editors*: Allah Bakhsh and Muhammad Rafiq Choudhry.

Ayers, R. S. and Westcot, D. W. 1994. Water Quality for Agriculture, Irrigation and Drainage Paper 29, rev. 1, Food and Agriculture Organization of the United Nations, Rome.

Bauder, T. A., Waskom, R. M., Sutherland, P. L. and Davis, J. G. Irrigation Water Quality Criteria Fact Sheet No. 0.506 Crop Series, Irrigation, Colorado State University.

CWT, 2004. Clean Water Team, State Water Resources Control Board FACT SHEET 3.1.5.9. Color of Water Fact Sheet. pp 1-3.

Fipps, G. Irrigation Water Quality Standards and Salinity Management Strategies. Texas A&M Extension Service. The Texas A&M University System. U S Department of Agriculture and The County Commissioners Courts of Texas Cooperating.

George, P. R. 1983. *Agricultural water quality criteria: irrigation aspects.* Department of Agriculture and Food, Western Australia. Report 30.

Kahlown, M. A. and Kahn, A. D. 2002. Irrigation Water Quality Manual. Pakistan Council of Research in Water Resources, Ministry of Science and Technology, Government of Pakistan, Islamabad.

Kahlown, M. A., Tahir, M. A., Rashid, H. and Bhatti, K. P. 2006. Water Quality Status. Fourth Technical Report 2004-06, PCRWR, Islamabad.

Mass. 1987. Salt tolerance of plants. *CRC Handbook of Plant Science in Agriculture.* B.R. Cristie (ed.). CRC Press Inc.

Miller, R. W. and Gardiner, D. T. 2007. Soils in our environment. 9th edition. Prentice Hall-Inc., Upper Sddle River, New Jersey 07458. ISBN 0-13-020036-0, Table 15-6, page 452.

Tanji, K. K. 1990. *Agricultural Salinity Assessment and Management.* American Society of Civil Engineers. Manuals and Reports on Engineering Practice Number 71. 619pp.

14

Efficient Utilization of Saline Water in Agriculture Irrigation

Swati Shabnam and Shushama Majhi

Introduction

One of the most common concerns in irrigated agriculture is regarding water quality or salinity. Out of the ground water surveyed in different Indian States, as much as 84% is rated either saline or alkaline (Minhas 1996). Salinity management is very important for using saline water in irrigation safely. This requires an understanding of how salts affect soil as well as plants, of how hydrogeologic processes affect salt accumulation, and also of how cropping and irrigation activities affect soil and water salinity.

The objectives of management practices for getting optimal crop production with saline irrigation water include the prevention of salt build-up in soil which limit the productivity of soils and the control of salt balances in the soil-water system, as well as minimising the damaging effects of salinity on crop growth (Meiri and Plaut 1985). Amongst the various soil, water, and crop management options available, there is usually no single way to achieve safe use of saline water in irrigation but many different approaches and practices can be combined into satisfactory irrigation systems using saline water. The appropriate combination depends upon edaphic, hydrogeologic, climatic factors as well as social and economic condition of the farmer. Several practices interact with each other, and should be considered in an integrated manner, along with climatic and human factors. Because of the climate, the basic principles of saline water management need some adaptation, e.g. providing for a leaching requirement is not appropriate when the growing season for post-monsoon winter crops starts with a surface-leached soil profile, because it would increase the salt load. High salinities during the initial stages of growth are particularly harmful. Recommendations for effective management of irrigation water salinity depend upon local soil properties, climate, and water quality; options of crops and rotations; and irrigation and farm management capabilities.

What is salinity?

All major irrigation water sources contain a variety of naturally occurring dissolved minerals which can vary with location, time, and source of water. Many of these mineral salts are beneficial being micronutrients. However, excessive total salt concentration or excessive levels of some potentially toxic elements can be detrimental to plant health and/or soil conditions.

The term "salinity" is used to describe the concentration of (ionic) salts, generally including calcium (Ca^{2+}), magnesium (Mg^{2+}), sodium (Na^{+}), potassium (K^{+}), chloride (Cl^{-}), bicarbonate (HCO_3^{-}), carbonate (CO_3^{2-}), sulfate (SO_4^{2-}) and others. Salinity, in general is expressed in terms of electrical conductivity (EC), in units of millimhos per centimeter (mmhos/cm), micromhos per centimeter (µmhos/cm), or deciSiemens per meter (dS/m). The electrical conductivity of a water sample is proportional to the concentration of the dissolved ions in the sample; hence EC is a simple indicator of total salt concentration.

Another frequently used term for describing water quality is Total Dissolved Solids (TDS), which is a measure of the mass concentration of dissolved constituents in water. TDS generally is reported in units of milligrams per liter (mg/l) or parts per million (ppm). Specific salts reported on a laboratory analysis report are also expressed in terms of mg/l or ppm and these represent mass concentration of each component in the water sample. Another term used to express mass concentration is normality; units of normality are milligram equivalents per liter (meq/l). The most common units used in expressing salinity are summarized in Table 14.1.

Why is salinity a problem?

Salinity in water (or soil solution) causes an osmotic potential. In simple terms, the salts in soil solution or in the soil compete with the plants for available water. Some salts can have a toxic effect on the plant or can burn plant roots and/or foliage. Some minerals, when present in excessive amount, may interfere with relative availability and uptake of other micronutrients by plants.

High concentration of sodium in soil can lead to the dispersion of soil aggregates damaging soil structure and interfering with soil permeability. Hence special consideration of the sodium level or sodicity in soils is warranted.

Saline waters as a resource

The lack of fresh water resources for irrigation is limiting the sustainable development of agriculture worldwide. In recent years saline water has been widely utilized as a supplemental irrigation water source in fresh water-deficient areas (Feng *et al.*, 2019). There is ample evidence to illustrate the potential of using saline waters for irrigation under different conditions. This evidence and experience demonstrate that waters of much higher salinities than those customarily classified as "unsuitable for irrigation" can be effectively used for the production of selected salt tolerant crops under the right conditions.

Table 14.1: Units commonly used to express salinity

Mass Concentration (Total Dissolved Solids)

mg/l = milligrams per liter ppm = parts per million ppm = mg/l

Electrical Conductivity (increases with increasing TPS)

conductivity = 1/resistance (mho = 1/ohm)
millimhos/cm = millimhos per centimeter
μhos/cm = micromhos per cm
dS/m = deciSiemens per meter
1 dS/M = 1 mmho/cm = 1000 pmho/cm

Salinity Conversions

0.35 X (EC mmhos/cm) = osmotic pressure in bars
651 X (EC mmhos/cm) = TDS in mg/l
10 X (EC mmhos/cm) = Normality in meq/l
0.065 X (EC mmhos/cm) = percent salt by weight

*** Also has been related as**

TDS (mg/l) = EC (dS/m) X 640 for EC < 5 dS/m
TDS (mg/l) = EC (dS/m) X 800 for EC > 5 dS/m

Normality

meq/l = milligram equivalents per liter (aka milliequivalents per liter)
meq/l = mg/l -5- equivalent weight
equivalent weight = atomic weight electrical charge

Example

To convert 227 ppm calcium concentration to meq/l:
ppm = mg/l; therefore 227 ppm = 227 mg/l
Calcium atomic weight = 40.078 g/mol
valence: +2 (charge = 2)
equivalent weight = 40.078 / 2 = 20.04
meq/l = 227/20.04 = 11.33
Therefore 227 mg/l = 11.33 meq/l for calcium.

*Compiled from various sources

Assessing the suitability of saline water for crop production

The suitability of a water for irrigation should be evaluated on the basis of criteria indicative of its potential to create soil conditions hazardous to crop growth. The primary criteria for judging irrigation water quality in terms of potential hazards to crop growth are:

- **Permeability and tilth**
- **Salinity**
- **Toxicity and nutritional imbalance**

Permeability and crusting hazards are evaluated by electrical conductivity (EC_{iw}) and the sodium adsorption ratio predicted to occur in the top soil after irrigation (SAR) with reference to threshold tolerances (permissible combinations of EC_i, and SAR,) established for the specific soil in question or, in the absence of specific information, an appropriate general relation. SAR is predicted using a computer model (such as Watsuit) or, in the absence of a computer model, using the SAR value of the irrigation water (SAR_{iw}) (Rhoades *et al.* 1992).

Salinity, toxicity and nutritional problems are evaluated by comparing levels of soil water salinity, concentrations of toxic ions and ratios of Ca/Mg predicted to result in the rootzone of the soil after irrigation with reference to acceptable values of salinity, toxic-ion concentrations and Ca/Mg ratios for the crop(s) in question. Acceptable levels of salt and toxic-ion concentrations for many crops are given in Table 14.2. Predictions of soil salinity resulting from irrigation with a given saline water can also be made without benefit of a computer by ignoring salt precipitation and dissolution reactions.

In considering the use of a saline water for irrigation and in selecting appropriate management to protect water quality, it is important to recognize that the total volume of a saline water supply cannot be beneficially consumed for irrigation and crop production; and the greater its salinity, the less it can be consumed before the salt concentration becomes limiting. The practice of blending or diluting excessively saline waters with good quality water supplies should only be undertaken after consideration is given to how this affects the volumes of consumable water in the combined and separate supplies. Blending or diluting drainage waters with good quality waters in order to increase water supplies or to meet discharge standards may be inappropriate under certain situations. More crop production can usually be achieved from the total water supply by keeping the water components separated. Serious consideration should be given to keeping saline drainage waters separate from the "good quality" water supplies, especially when the latter waters are to be used for irrigation of salt-sensitive crops. The saline drainage waters can be used more effectively by substituting them for "good quality" water to irrigate certain crops grown in the rotation after seedling establishment.

In any case, it is imperative that management practices for the control of soil and water salinity at such scales be considered an essential part of the management requirements for using saline waters for irrigation. This requires the following:

- Seriousness of salinity-related environmental problems and the vulnerability of irrigated lands to waterlogging and salination be sufficiently recognized;

- Processes contributing to these problems and the effects of salts on soils and plants be understood;
- Salinity conditions and trends of the irrigated lands and associated water resources be routinely assessed using appropriate measurement and monitoring techniques that provide meaningful and timely information;
- Salinity-related problems be properly diagnosed using appropriate criteria and standards;
- Future conditions of soil and water salinity be adequately predicted using appropriate prognostic techniques; and
- Viability of the irrigated agriculture and associated water resources be sustained by implementing effective long-term control measures.

Table 14.2: Relative salt tolerance of various crops at emergence and during growth to maturity

Crop		Electrical conductivity of saturated soil extract	
Common name	Botanical name[1]	50% yield dS/m	50% emergence[2] dS/m
Barley	*Hordeum vulgare*	18	16-24
Cotton	*Gossypium hirsutum*	17	15
Sugarbeet	*Beta vulgaris*	15	6-12
Sorghum	*Sorghum bicolor*	15	13
Safflower	*Carthamus tinctorius*	14	12
Wheat	*Triticum aestivum*	13	14-16
Beet, red	*Beta vulgaris*	9.6	13.8
Cowpea	*Vigna unguiculata*	9.1	16
Alfalfa	*Medic ago sativa*	8.9	8-13
Tomato	*L ycopersicon lycopersicum*	7.6	7.6
Cabbage	*Brassica oleracea capitata*	7.0	13
Maize	*Zea mays*	5.9	21-24
Lettuce	*Lactuca sativa*	5.2	11
Onion	*Allium cepa*	4.3	5.6-7.5
Rice	*Oryza sativa*	3.6	18
Bean	*Phaseo/us vulgaris*	3.6	8.0

[1]Botanical and common names follow the convention of Hortus Third where possible.
[2]Emergence percentage of saline treatments determined when non-saline treatments attained maximum emergence.

Quality characteristics of saline waters

Chemical and physical characteristics of irrigation waters are restricted to those which predominantly affect crop production either directly or indirectly. The limiting values of the quality parameters vary considerably depending upon circumstances of use. Sewage and industrial effluents are not considered as the focus of these guidelines is on irrigation with drainage waters and moderately saline natural waters of various kinds. An abbreviated classification of waters in terms of salinity is given in Table 14.3 to facilitate the identification of the kinds of saline waters included in the scope of these guidelines.

Definitions and indices of salinity related parameters

The term salinity used herein refers to the total dissolved concentration of major inorganic ions (i.e. Na, Ca, Mg, K, HCO_3, SO_4 and Cl) in irrigation, drainage and groundwaters. Individual concentrations of these cations and anions in a unit volume of the water can be expressed either on a chemical equivalent basis, mmol/l, or on a mass basis, mg/l. Total salt concentration (i.e. salinity) is then expressed either in terms of the sum of either the cations or anions, in mmol/l, or the sum of cations plus anions, in mg/l. For reasons of analytical convenience, a practical index of salinity is electrical conductivity (EC), expressed in units of deciSiemen per metre (dS/m). An approximate relation (because it also depends upon specific ionic composition) between EC and total salt concentration is 1 dS/m = 10 mmol/l = 700 mg/l. Electrical conductivity values are always expressed at a standard temperature of 25°C to enable comparison of readings taken under varying climatic conditions. With all its obvious shortcomings, this custom of using EC as an index of salinity emphasizes the concept that, as a good first approximation, plants respond primarily to total concentration of salts rather than to the concentrations or proportions of individual salt constituents.

A similar usage of EC for expressing soil salinity has evolved, where the parameter of primary interest is the total sail concentration, or EC, of the soil solution. However, the content of water in the soil is not constant over time nor is the composition of the soil solution. For this reason, soil salinity is not an easily defined, single-valued parameter. In an attempt to standardize measurements and to establish a reasonable reference for comparison purposes, "soil salinity", is commonly expressed in terms of the electrical conductivity of an extract of a saturated paste (ECe; in dS/m) made using a sample of the soil.

In addition to total salt concentration, sodium and pH can adversely affect soil properties for irrigation and cropping. At high levels of sodium relative to divalent cations in the soil solution, clay minerals in soils tend to swell and disperse and aggregates tend to slake, especially under conditions of low total salt concentration and high pH. Whether from slaking, swelling or from clay dispersion, the permeability of the soil is reduced and the surface becomes more crusted and compacted under such conditions. Thus, the ability of the soil to transmit water can be severely reduced by excessive sodicity (the term used herein to refer to the combined deleterious effects of high sodium and pH, and low electrolyte concentration on soil physical properties). Since high total salt concentration tends to increase a soil's stability with respect to aggregation and permeability, distinction is made between saline soils and sodic soils. With respect to sodicity, it is the proportion of adsorbed exchangeable sodium relative to the cation exchange capacity (often expressed as the exchangeable sodium percentage, ESP), rather than the absolute amount of exchangeable sodium, that is relevant along with the total salt concentration of the infiltrating and percolating water and the soil pH. Because ESP and the sodium adsorption ratio of the saturation extract (SAR = NAtV(Ca+Mg)/2, where solute concentrations are in mmole/l) are so closely related, SAR is commonly used as a substitine for ESP and as an index of the sodium hazard of soils and waters.

Certain ions in saline waters can be specifically toxic to plants, if present in excessive concentrations or proportions. Of particular concern are sodium (Na), chloride (Cl), and boron (B). While not often toxic to plants, a few solutes sometimes (though not frequently) found in natural saline waters may accumulate in plant parts at levels that can be toxic to consumers, if their diet is largely restricted to this food. Such elements include selenium (Se), arsenic (As), and molybdenum (Mo).

Classification of saline waters

Because the suitability of saline water for irrigation is so dependent upon the conditions of use, including crop, climate, soil, irrigation method and management practices, water quality classifications are not advised for assessing water suitability for irrigation. However, for the purpose of identifying the levels of water salinities for which these guidelines are intended, it is useful to give a classification scheme.

Such a classification is given in Table 14.3 in terms of total salt concentration, which is the major quality factor generally limiting the use of saline waters for crop production. Only very tolerant crops (hardly any conventional crops) can be successfully produced with waters that exceed about 10 dS/m in EC. Few generally-used irrigation waters exceed about 2 dS/m in EC. Many drainage waters, including shallow groundwaters underlying irrigated lands, fall in the range of 2-10 dS/m in EC. Such waters are in ample supply in many developed irrigated lands and have good potential for selected crop production, though they are often not used in this regard and are more typically discharged to better quality surface waters or to waste outlets. Reuse of second-general ion drainage waters for irrigation is also sometimes possible and useful, especially for the purpose of reducing drainage volume in preparation for ultimate disposal or treatment. Such waters will generally have ECs in the range 10-25 dS/m. Thus, they too are considered in these guidelines, though to a much lesser degree because the "crops" that can be grown with them are atypical and much less experience exists upon which to base management recommendations and to develop guidelines. Very highly saline waters (25 - 45 dS/m in EC) and brine (> 45 dS/m in EC) are beyond the scope of these guidelines and their uses for crop production are therefore not discussed herein.

Table 14.3: Classification of saline waters

Water class	Electrical conductivity ds/m	Salt concentration mg/l	Type of water
Non saline	< 0.7	<500	Drinking and irrigation water
Slightly saline	0.7-2	500-1500	Irrigation water
Moderately saline	2-10	1500-7000	Primary drainage water and ground water
Highly saline	10-25	7000-15000	Secondary drainage water and ground water
Very highly saline	25-45	15000-35000	Very saline ground water
Brine	>45	>35000	Sea water

Sources and availability of saline waters

In practical agricultural use, a common source of saline water is groundwater. Salinity of groundwater can be man-induced or natural. In many areas, saline and fresh subsurface waters exist in close proximity. When fresh groundwater is pumped from aquifers that are in hydraulic connection with seawater, the change in gradients as a result of pumping may result in a flow of salt water from the sea towards the well. This is called seawater intrusion.

Upcoming is another mechanism by which groundwater could become brackish. Upcoming refers to a situation where a well, located close enough to saline water underlying freshwater, is pumped at a rate sufficient to cause the salt water to be drawn into the well in an upward shaped cone or mound.

There are also natural causes of salinity. Numerous investigators have noted that water within sedimentary strata becomes increasingly saline with an increase in depth. In general, the sequence noted is sulphate-rich water near the surface, saline bicarbonate water at an intermediary level and more concentrated chloride water at greater depths (Craig 1980). There are several mechanisms by which water trapped in sedimentary rocks can be altered into saline water. One of these is the solution of sediments and rocks.

In coastal regions, surface water sources can become saline due to the tidal influence of the sea. As the high tide moves into the coastal area, seawater moves into streams and drainage canals and travels inland. This upstream migration of seawater alters the quality of water in affected streams and drainage canals significantly. This phenomenon is also observed during times of drought.

Another important source of saline water is drainage effluent (including perched groundwater) from irrigated areas. Drainage water, once thought of as wastewater, is now used in many countries for irrigation. The salinity levels vary, but often the salt levels are higher than those of conventional primary irrigation water sources. Reuse of drainage effluent is important when the supply of good quality irrigation water is limited, and it is also an efficient means of reducing water pollution.

Gupta (1990) has treated the subject of saline water use in India comprehensively. He reported that the salinity level of the Ganges river in India is very low and average total dissolved salt concentration is less than 200 mg/l. However, there are specific stretches or locations along the river system where salinity level increases due to hydrologic as well as human-induced activities. In the deltaic region of the Ganges river in West Bengal, which comes under tidal influences, the salinity of the water can rise to greater than 10 times the average salinity of the river.

In the Punjab, Maharashtra area, canal waters are reported to be of good quality with EC values often less than 0.5 dS/m. On the other hand, drainage waters are reported to have high salinities. Prasad (1967) reported that the drainage waters of the Unnao Tehsil in Uttar Pradesh had an average EC of 2 dS/m. Gupta (1990) carried out a survey of groundwater quality in Rajasthan and estimated percentages of wells that fall into varying classes of salinity. The results are presented in Table 14.4.

Table 14.4: Percent distribution of groundwaters of Rajasthan in different EC classes

EC range dS/m	11 arid districts (2817)[1]	7 semi-arid districts (4000)[1]	8 humid districts (2614)[1]
<0.75	10	23	41
0.75-2.25	29	48	49
2.25-5.00	27	19	8
5.00-10.0	20	8	2
10.0-15.0	9	2	-
>15.0	5	-	-

1= Number of samples

Management Principles

To prevent the excessive accumulation of salt in the rootzone from irrigation, extra water (or rainfall) must, over the long term, be applied in excess of that needed for ET and must pass through the rootzone in a minimum net amount. This amount, in fractional terms, is referred to as the leaching requirement (Lr, the fraction of infiltrated water that must pass through the rootzone to keep salinity within acceptable levels; US Salinity Laboratory Staff 1954). In fields irrigated to steady-state conditions with conventional irrigation management, the salt concentration of the soil water is essentially uniform near the soil surface regardless of the leaching fraction (Lf, the fraction of infiltrated water that actually passes through the root-zone) but increases with depth as Lf decreases. Likewise, average rootzone salinity increases as LF decreases; crop yield is decreased when tolerable levels of salinity are exceeded. Methods to calculate the leaching requirement and to predict crop yield losses due to salinity effects were described previously. Once the soil solution has reached the maximum salinity level compatible with the cropping system, at least as much salt as is brought in with additional irrigations must be removed from the rootzone; a process called "maintaining salt balance."

To prevent waterlogging and secondary salination, drainage must remove the precipitation and irrigation water infiltrated into the soil that is in excess of crop demand and any other excessive water (surface or subsurface) that flows into the area; it must provide an outlet for the removal of salts that accumulate in the rootzone in order to avoid excessive soil salinization, and it must keep the water table sufficiently deep to permit adequate root development, to prevent the net flow of salt-laden groundwater up into the rootzone by capillary forces and to permit the movement and operation of farm implements in the fields. Artificial drainage systems may be used in the absence of adequate natural drainage. They are essentially engineering structures that control the water table at a safe level according to the principles of soil physics and hydraulics. The water table depth required to prevent a net upward flow of water and salt into the rootzone is dependent on irrigation management and is not single-valued as is commonly assumed (van Schilfgaarde 1976).

Management practices for the safe use of saline water for irrigation primarily consist of:

- Selection of crops or crop varieties that will produce satisfactory yields under the existing or predicted conditions of salinity or sodicity;
- Special planting procedures that minimize or compensate for salt accumulation in the vicinity of the seed;
- Irrigation to maintain a relatively high level of soil moisture and to achieve periodic leaching of the soil;
- Use of land preparation to increase the uniformity of water distribution and infiltration, leaching and removal of salinity;
- Special treatments (such as tillage and additions of chemical amendments, organic matter and growing green manure crops) to maintain soil

permeability and tilth. The crop grown, the quality of water used for irrigation, the rainfall pattern and climate, and the soil properties determine to a large degree the kind and extent of management practices needed.

Growing suitably tolerant crops

Where salinity cannot be kept within acceptable limits by leaching, crops should be selected that can produce satisfactory yields under the resulting saline conditions. In selecting crops for saline soils, particular attention should be given to the salt tolerance of the crop during seedling development, because poor yields frequently result from failure to obtain a satisfactory stand. Some crops that are salt tolerant during later stages of growth are quite sensitive to salinity during early growth. Tolerances of the various major crops to salinity are given in Table 14.2.

Managing seedbeds and grading fields to minimize local accumulations of salinity

Failure to obtain a satisfactory stand of furrow-irrigated row crops on moderately saline soils is a serious problem in many places. This is because the rate of germination is reduced by excessive salinity, as previously discussed. The failures are usually due to the accumulation of soluble salt in raised beds that are "wet-up" by furrow irrigation. Modifications in irrigation practice and bed shape should be used to reduce salt accumulation near the seed. The tendency of salts to accumulate near the seed during irrigation is greatest in single-row, round-topped beds (Figure 14.1).

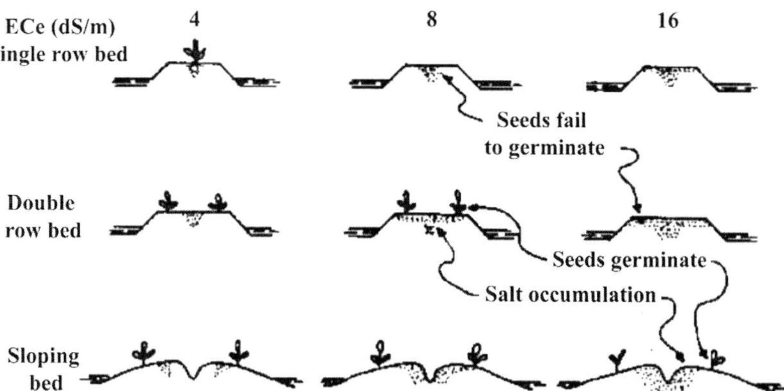

Fig. 14.1: Pattern of salt build-up as a function of seed placement, bedshape and level of soil salinity (after Bernstein, Fireman and Reeve 1955)

Sufficient salt to prevent germination may concentrate in the seed zone even if the average salt content of the soil is moderately low. Thus, such beds should be avoided when irrigating with saline waters using furrow methods. With double-row, flat-topped beds, since most of the salt moves into the centre of the bed, the shoulders are left relatively free of salt, thus seedling establishment may be enhanced by planting on the shoulders of such beds. Sloping beds are best for saline conditions because the seed can be safely planted on the slope below the zone of high salt accumulation. Such beds should be used, if possible, when furrow irrigating with saline waters. Planting in furrows or basins is satisfactory from the standpoint of salinity control but is often unfavourable for the emergence of many row crops because of problems related to crusting and poor aeration. This method is recommended only for the use of very saline irrigation waters and vigorous, hardy emerging plants. Pre-emergence irrigation by use of sprinklers or special furrows placed close to the seed may be used to keep the soluble salt concentration low in the seedbed during germination and seedling establishment. After the seedlings are established, the special furrows may then be abandoned and new furrows made between the rows, and sprinkling replaced by furrow irrigation.

Careful grading of land makes possible a more uniform application of water and, hence, better salinity control when irrigating with saline water. Barren or poor areas in otherwise productive fields are often either high spots that do not receive enough water for good crop growth or for leaching purposes or low spots that remain too wet for seedling establishment. Lands that have been irrigated one or two years after initial grading usually need to be regraded to remove the surface unevenness caused by the settling of fill material. Annual crops should be grown after the first grading so that regrading can be performed before a perennial crop is planted. A prior detailed topographic survey could be very helpful to avoid ruining soil properties and in particular removing the surface soil which may be relatively more fertile. Land levelling causes a significant soil compaction due to the weight of the heavy equipment and it is advisable to follow this operation with subsoiling, chiselling and ploughing to break up the compaction and restore or improve water infiltration (Rhoades *et al.* 1992).

Managing soils under saline water irrigation

Several physical, chemical and biological soil management measures help facilitate the safe use of saline water in crop production. Some important ones in this regard are: tillage, deep ploughing, sanding, use of chemical amendments and soil conditioners, organic and green manuring and mulching.

Tillage is a mechanical operation that is usually carried out for seedbed preparation, soil permeability improvement, to break up surface crusts and to improve water infiltration. If tillage is improperly executed, it might form a plough layer or bring a salty layer closer to the surface. Sodic soils are especially subject to puddling and crusting; they should be tilled carefully and wet soil conditions avoided. Heavy machinery traffic should also be avoided. More frequent irrigation, especially during the germination and seedling stages, tends to soften surface crusts on sodic soils and encourages better stands.

Deep ploughing refers to depths of ploughing from about 40 to 150 cm. It is most beneficial on stratified soils having impermeable layers lying between permeable layers. In sodic soils, deep ploughing should be carried out after removing and reclaiming the sodicity, otherwise it will cause complete disturbances and collapse of the soil structure. Deep ploughing to 60 cm loosens the aggregates, improves the physical condition of these layers, increases soil-water storage capacity and helps control salt accumulation when using saline water for irrigation. Crop yields can be markedly improved by ploughing to this depth every three or four years. The selection of the right plough types (shape and spacings between shanks), sequence, ploughing depth and moisture content at the time of ploughing should provide good soil tilth and improve soil structure (Mashali 1989). Special equipment can even invert whole soil profiles or break up substrata as deep as 2.5m that impede deep percolation, so that many adverse physical soil conditions associated with land irrigated with saline water can be modified in order to improve leachability and drainability.

Sanding is used in some cases to make a fine textured surface soil more permeable by mixing sand into it, thus a relatively permanent change in surface soil texture is obtained. When properly done, sanding results in improved root penetration and better air and water permeability which facilitates leaching by saline sodic water and when surface infiltration limits water penetration. The method can be combined with initial deep ploughing.

Chemical amendments are used to neutralize soil reaction, to react with calcium carbonate and to replace exchangeable sodium by calcium. This decreases the ESP and should be followed by leaching for removal of salts derived from the reaction of the amendments with sodic soils. They also decrease the SAR of irrigation water if added in the irrigation system. Gypsum is by far the most common amendment for sodic soil reclamation, particularly when using saline water with a high SAR value for irrigation. Calcium chloride is highly soluble and would be a satisfactory amendment especially when added to irrigation water. Lime is not an effective amendment for improving sodic conditions when used alone but when combined with a large amount of organic manure it has a beneficial effect. Sulphur too can be effective; it is inert until it is oxidized to sulphuric acid by soil micro-organisms. Other sulphur-containing amendments (sulphuric acid, iron sulphate, aluminium sulphate) are similarly effective because of the sulphuric acid originally present or formed upon microbial oxidation or hydrolysis.

The choice of an amendment for a particular situation will depend upon its relative effectiveness judged from its improvement of soil properties and crop growth, the availability of the amendments, relative cost involved, handling and application difficulties and time allowed and required for the amendment to react in soil and effectively replace adsorbed sodium.

Mineral fertilizers: Salt accumulation affects nutrient content and availability for plants in one or more of the following ways: by changing the form in which the nutrients are present in the soil; by enhancing loss of nutrients from the soil through heavy leaching or, as in nitrogen, through denitrification, or by precipitation in soil; through the effects of non-nutrient (complementary) ions on nutrient uptake; and by adverse interactions between the salt present in saline water and fertilizers, decreasing fertilizer use efficiency.

Crop response to fertilizer under saline or sodic conditions is complex since it is influenced by many soil, crop and environmental factors. The benefits expected from using soil management measures to facilitate the safe use of saline water for irrigation will not be realized unless adequate, but not excessive, plant nutrients are applied as fertilizers. The level of salinity may itself be altered by excess fertilizer application as mineral fertilizers are for the most part soluble salts. The type of fertilizer applied, when using saline water for irrigation, should preferably be acid and contain Ca rather than Na taking into consideration the complementary anions present. Timing and placement of mineral fertilizers are important and unless properly applied they may contribute to or cause a salinity problem.

Organic and green manures and mulching: Incorporating organic matter into the soil has two principal beneficial effects of soils irrigated with saline water with high SAR and on saline sodic soils: improvement of soil permeability and release of carbon dioxide and certain organic acids during decomposition. This will help in lowering soil pH, releasing calcium by solubilization of $CaCO_3$ and other minerals, thereby increasing ECe and replacement of exchangeable Na by Ca and Mg which lowers the ESP. Growing legumes and using green manure will improve soil structure. Green manure has a similar effect to organic manure. Salinization during fallowing may be severe where a shallow water table exists, since evaporation rates of about 8, 3 and 1 mm/day could occur from the dry surface of fine sandy loam when the water table is kept at 90, 120 and 180 cm from the soil surface, respectively. Mulching to reduce evaporation losses will also decrease the opportunity for soil salinization. When using saline water where the concentration of soluble salts in the soil is expected to be high in the surface, mulching can considerably help leach salts, reduce ESP and thus facilitate the production of tolerant crops. Thus, whenever feasible, mulching to reduce the upward flux of soluble salts should be encouraged.

Operating delivery systems efficiently

Water delivery and distribution systems must be operated efficiently to facilitate the timely supply of water in the right quantities and to avoid waterlogging and salinity build-up in irrigated lands, especially when saline waters are involved. The amount of water applied should be sufficient to supply the crop and satisfy the leaching requirement but not enough to overload the drainage system. Over-irrigation contributes to the high water table, increases the drainage requirement and is a major cause of salinity build-up in many irrigation projects of the world. Therefore, a proper relation between irrigation, leaching, and drainage must be maintained in order to prevent irrigated lands from becoming excessively waterlogged and salt-affected.

Excessive loss of irrigation water from canals constructed in permeable soil is a major cause of high water tables and secondary salination in many irrigation projects. Such seepage losses should be reduced by lining the canals with impermeable materials or by compacting the soil to achieve a very low permeability. Because the amount of water passing critical points in the irrigation delivery system must be known in order to provide water control and to achieve high water-use efficiency, provisions for effective flow measurement should be made. Unfortunately, many current irrigation systems do not use flow measuring devices and, thus, the farmers operate with limited control and knowledge of the amount of water actually diverted to the farms. In addition, many delivery systems encourage over-irrigation because water is supplied for fixed periods, or in fixed amounts, irrespective of seasonal variations in on-farm needs. Salinity and water table problems are often the result. The distribution system should be designed and operated so as to provide water on demand and in metered amounts as needed to achieve high efficiency and to facilitate salinity control and the use of saline waters for irrigation.

Irrigating efficiently

Improvements in salinity control generally come hand-in-hand with improvements in irrigation efficiency. The key to the effective use of saline irrigation waters and salinity control is to provide the proper amount of water to the plant at the proper time. The ideal irrigation scheme should provide water as nearly continuously as possible, though not in excess, as needed to keep the soil water content in the rootzone within optimum safe limits. However, carefully programmed periods of stress may be needed to obtain maximum economic yield with some crops; cultural practices also may demand occasional periods of dry soil. Thus, the timing and amount of water applied to the rootzone should be carefully controlled to obtain good water use efficiency and good crop yield, especially when irrigating with saline water. As mentioned above, this requires water delivery to the field on demand which, in turn, requires the establishment of close coordination between the farmer and the entity that distributes the water; it calls for the use of feedback devices to measure the water and salt contents and potentials in the soil and devices to measure water flow (rates and volumes) in the conveyance systems.

The method and frequency of irrigation and the amount of irrigation water applied may be managed to control salinity. The main ways to apply water are basin flooding, furrow irrigation, sprinkling, subirrigation, and drip irrigation. Flood irrigation is good for salinity control when using saline waters if the land is level, though aeration and crusting problems may occur. Aeration and crusting problems are minimized by using furrow irrigation, but salts tend to accumulate in the beds. If excess salt does accumulate, a rotation of crops and periodic irrigation by sprinkler or flooding should be used as salinity-control measures. Alternatively, cultivation and irrigation depths should be modified, once the seedlings are well established, to "shallow" the furrows so that the beds will be leached by later irrigations. Irrigation by sprinkling may give better control of the amount and distribution of water; however, the tendency is to apply too little water by this method, and leaching of salts beyond the rootzone may sometimes be accomplished only with special effort. Salinity can be kept low in the seedbed during germination with sprinkler-irrigation, but crusting may be a problem. Emergence problems associated with such crusting may be overcome with frequent light irrigations during this time or by use of special tillage techniques. Sub irrigation with saline water is not generally advisable unless the soil is periodically leached of the accumulated salts by rainfall or by surface applications of low-salinity water. Drip irrigation, if properly designed, is recommended for use of saline irrigation water because it minimizes salinity and matric stresses in the rootzone, though salts accumulate in the periphery of the wetted area. As noted earlier, higher levels of salinity in the irrigation water can be tolerated with drip as compared with other methods of irrigation.

Because soluble salts reduce the availability of water in almost direct proportion to their total concentration in the soil solution, irrigation frequency should be increased so that the moisture content and salinity of irrigated soils are maintained as high and low, respectively, as is practicable, especially during seedling establishment and the early stage of vegetative growth, if it can be done without resulting in excessive leaching or insufficient depth of rooting. The most practical way to accomplish this is through use of drip irrigation.

Additional water (over that required to replenish losses by plant transpiration and evaporation) must be applied, at least occasionally, to leach out the salt that has accumulated during previous irrigations. This leaching requirement depends on the salt content of the irrigation water and on the maximum salt concentration permissible in the soil solution which depends in turn on the salt tolerance of the crop and the manner of irrigation. The first irrigations provided for the renewal of cropping following a fallow or uncropped period often unavoidably result in relatively high leaching. Many irrigation practices, especially with flood irrigation, inadvertently result in excess leaching, especially during pre-plant or early-season irrigations before the soil aggregation has slaked and surface soil permeability has diminished. Effects of non-uniform crop stand and cover, soil infiltration rates (permeabilities) and water application and distribution result in generally non-uniform leaching across an irrigated field. Calculation of the leaching requirement is disproportionately subject to errors related to uncertainties in knowledge of evapotranspiration, since Lr, = 1 - Vet/Viw. The value, much less the distribution, of evapotranspiration is not precisely known for most field situations, especially for conditions of irrigation with saline waters and in the presence of shallow, saline water tables. Consequently, there is little documented evidence of the positive benefits of increased leaching on crop yield under actual field conditions when irrigating with saline waters (Shalhevet 1984). While, certainly, the excess salts applied with saline irrigation waters must be removed over time to sustain crop production, for both short- and long-season crops it is generally sufficient to intentionally apply extra water for leaching only if and when the levels of salinity in the active rootzone actually become excessive. Giving extra water for leaching according to traditional Lr,. equations with each and every irrigation is not necessary. Rainfall in sub-humid climates often provides the required leaching.

In furrow-irrigated areas, furrow length should be reduced in order to improve intake distribution and to reduce tail water runoff. Surge irrigation techniques can sometimes be used to improve irrigation uniformity in graded furrows (Bishop *et al.* 1981). For tree crops, a low-head bubbler system can be used to provide excellent control and to minimize the pressure requirements and expensive filtration systems (Rawlins 1977). Drip systems, of course, are increasingly being used for permanent crops and high-value annual crops and are well suited for use with saline irrigation waters. All opportunities to modify existing irrigation systems to increase their effectiveness of water and salinity control should be sought and implemented.

A frequent constraint in improving on-farm water use is the lack of knowledge of just when irrigation is needed and of how much capacity for storage is available in the rootzone. Ways to detect the onset of plant stress and to determine the amount of depleted soil water are prerequisites to supplying water on demand and in the amount needed. Prevalent methods of scheduling irrigation usually do not, but should, incorporate salinity effects on soil-water availability (Rhoades *et al.* 1981). When irrigating with saline waters, the osmotic component of the soil water potential of the rootzone must be considered in scheduling decisions.

Ideally, irrigation management should have the available soil water near the upper limit during germination and emergence but depleted by about 50 percent, or more, at harvest and should maintain available water within the major rootzone during the early vegetative, flowering and yield formation growth stages at a level which produces no deleterious plant water stress through successive, properly-timed irrigations (FAO 1979). Under saline conditions, some "extra" water must be given for leaching - a minimum commensurate with salt tolerance of the crop being grown, if rainfall is inadequate in this regard, as discussed previously. Some method of assessing the water availability to the crop with sufficient lead time to provide for a water application before significant stress occurs should be used for irrigation scheduling purposes. In addition, the amounts of water needed for replenishment of the depleted soil moisture from the rootzone and for leaching must be determined.

Various scheduling methods can be used which are based on sensing depletion of soil -water per se or soil water potential (matric, osmotic or total), or some associated soil or water property, and knowledge of the critical level (the set-point value). Such levels can be ascertained from salt tolerance data by converting threshold ECe values to osmotic potentials and assuming equivalent crop yield loss (also ET loss) would result from total water potential (i.e. assuming the effects of matric and osmotic potentials are equivalent and additive). Matric potential should be measured by any suitable means. Osmotic potential should be determined by one of the methods of salinity measurement and daily potential evapotranspiration can be calculated from measurements of air temperature, humidity, solar radiation and wind or of pan evaporation. The actual evapotranspiration (ETa) can then be estimated from empirically determined, crop coefficients. The summation of these daily ETa values can then be used to estimate accumulative soil water depletion and total water potential. A plot of depletion or water potential versus time is then used to project the need for irrigation. This basic approach can be used based regardless of whether direct measurements of soil water content, or a related parameter, using neutron meters, resistance blocks, time-domain reflectometric (TDR) sensors, four-electrode sensors, or various soil matric potential sensors, etc., are used or estimated from ET methods. All of the methods suffer the limitation of needing to know the critical set-point value for irrigation, which varies with crop type, rooting characteristics, stage of plant growth, soil properties and climatic stress, etc.

For saline water, irrigations should be scheduled before the total soil water potential (matric plus osmotic) drops below the level (as estimated above) which permits the crop to extract sufficient water to sustain its physiologic processes without loss in yield. Since, typically, the crop's root system normally extracts progressively less water with increasing soil depth (because rooting density decreases with depth and salt concentration increases with depth, as discussed earlier), the frequency of irrigations should be determined by the level of total soil water potential in the upper half of the rootzone where the rate of water depletion is greatest. Besides the extent of soil water depletion by ET, determination of the amount of water to apply should also be based on stage of plant development, the salt tolerance of the crop at this stage and the status of the soil water salinity at deeper depths in the rootzone. In early stages of plant development it is often desirable to irrigate just sufficiently to bring the soil to "field capacity" to the depth of present-rooting or just beyond. Eventually, however, excess water must be applied to leach salts accumulated in the upper profile to deeper depth in order to provide the growing plant access to more "usable" soil water in accordance with its expanding needs. Thus, the amount of irrigation water required is dictated by the plant's need for water, the volume of soil reservoir in need of replenishment and the level of soil salinity in the lower rootzone. Benefits of different amounts of saline irrigation water should be determined by evaluating their effects on relative crop yield using the water production function model.

Monitoring soil water and salinity and assessing adequacy of leaching and drainage

"Feedback" information on the status of salt and water within the crop rootzone and the extent of leaching being achieved should be obtained periodically to identify developing problem areas, to evaluate the appropriateness of model predictions and as a guide to monitor the effectiveness of the irrigation system and management strategies being used. Soil water content (or matric potential), salinity (and hence osmotic potential) and leaching fraction can, in theory, all be determined from measurements of soil electrical conductivity, ECa, since ECa is a measure of both soil water content and soil water salinity. Soil salinity in irrigated agriculture is normally low at shallow soil depths and increases through the rootzone. Thus measurements of EC, in shallow depths of the soil profile made over an irrigation cycle are relatively more indicative of changing soil water content (permitting estimation of matric potential), while measurements of ECa in deeper depths of the profile, where less water uptake occurs, are relatively more indicative of salinity. Thus, in principle, depletion of soil water to a set-point level, depth of water penetration from an irrigation or rainfall event and leaching fraction can all be determined from ECa measurements made within the rootzone over time. However, separate measurements of soil water content and soil water salinity, from which the total water potential can be estimated (matric plus osmotic), are more ideally suited for these needs. The use of time domain reflectometric (TDR) sensors offer potential in this regard (Dalton and Poss 1990).

Proper operation of a viable, permanent irrigated agriculture, which also uses water efficiently, requires periodic information on soil salinity, especially with use of saline waters. Only with this information can the effectiveness of irrigation project operation be assessed with respect to the adequacy of leaching and drainage, salt balance and water use efficiency. Monitoring programs should be implemented to evaluate the appropriateness of model predictions, the effectiveness of control programs, and to assess the adequacy of the irrigation and drainage systems on a project-wide basis. Frequently used methods based on "salt-balance" calculations are inadequate in this regard.

Examples of use of saline waters for irrigation

A selected review of some representative examples of the commercial use that has been made of saline waters for irrigation ur der different circumstances around the world follows. The examples were chosen to be representative of the worldwide experience of such use and because relevant information, including water quality, climate, soil type, crops, irrigation systems and methods, other management practices, yields and period of use, was available. They also illustrate some of the management practices that have been found to be effective to facilitate such use.

Crops are successfully grown in some parts of India under conditions quite different from those existing in typical, semi-arid regions. Much of the research and experience in India through 1980 has been summarized by Gupta and Pahwa (1981). Of particular benefit to the continued use of saline waters for irrigation in parts of India are the monsoon rains. It has been observed that very saline waters can be used for irrigation in these areas without excessive long-term build-up of soil salinity because of the extensive seasonal leaching that occurs there These findings illustrate the high potential to gain benefit from the use of quite saline waters for irrigation in regions which receive sufficient rainfall to prevent the build-up of excessive soil salinity over time.

A field survey made during the period 1983-1985 showed that extensive use (104 000 shallow tubewells pumping 106 000 hectare-metres of water per year) is being made (since about 1975) of shallow-saline groundwater of EC up to 8 dS/m for irrigation in nine districts of Haryana State India (Boumans *et al.* 1988). In four of the districts, the saline water is solely used for irrigation, while in the remaining five it is used either after it is blended with fresh canal water or in alternation with the canal water. Mean rainfall in these areas ranges between 300 and 1100 mm. The soils are dominantly sandy loam in texture. Shallow water tables exist and surface flooding occurs following the monsoons. Only a few wells had EC values exceeding 7 dS/m, hence it appears that this level is about the maximum that the farmers have found to be acceptable for long-term use. Yield depressions of 30-40 percent are apparently acceptable to these farmers. The farming practices being used were not given, so it is not possible to evaluate whether opportunities may exist to improve yields through the adoption of modified practices. Still it is obvious that saline waters have been used successfully, even as the sole supply, for irrigation in these districts of India. Whether their and vegetables. In the Delta, saline waters of EC 2.5 to 4 dS/m has been used successfully to grow vegetables under greenhouse conditions. In the New Valley (Oases, Siwa, Bahariya, Farafra, Dakhla and Kharga) there is potential to irrigate about 60 000 ha utilizing groundwater (salinity ranging from EC 0.5 dS/m to 6.0 dS/m), of which 17 000 ha are already under cultivation. Siwa Oasis has the largest naturally flowing springs in the New Valley. Siwa once contained a thousand springs, of salinity ranging from EC 2 to 4 dS/m, which were used successfully to irrigate olive and date-palm orchards, with some scattered forage areas. At present 3600 ha are irrigated from about 1200 wells. Of these 1000 are hand dug to depths of 20-25 m (salinity ranging from EC 3.5 to 5.0 dS/m and in some locations as much as 10 dS/m), and the remaining 200 wells were drilled deep (70-130 m) with salinity of EC 2.5-3.0 dS/m the SAR values varying from 5 to 20. Presently about 235 MCM/year is being used successfully to irrigate olive and date-palm orchards, alfalfa, cereals and wood trees (of which 60 MCM from continuing flowing springs). Due to over-irrigation without appropriate drainage facilities, seepage as well as run off to low lying land, salinity and waterlogging have developed in some lands of the oasis.

To reduce drainage water volumes, minimize water pollution and safely dispose of the ultimate unusable final drainage water, new strategies are being developed and experimented by the Government authorities in Siwa Oasis (similar problems exist in Dakhla oasis). These include:

- Use of natural flowing springs to irrigate winter crops such as cereals and forage;

- Use of saline water over 5 dS/m to irrigate salt tolerant crops like barley, vetches, Rhodes grass, sugarbeet, etc.;
- Use of biologically-active drainage water for the production of windbreak and growing wood trees;
- Use of drainage water for stabilization of sand dunes;
- Reuse of drainage water (average salinity is EC 6.0 dS/m with SAR values of 10 to 15) after blending with good quality water (recently drilled deep well of salinity EC 0.4 dS/m with SAR of 5) or by alternating the drainage water with good water.

References

Bernstein, L., Fireman, M. and Reeve, R.C. 1955. Control of salinity in the Imperial Valley, California. US Dept. Agric. ARS-41-4. pp.16.

Bishop, A.A., Walker, W.R., Allen, N.L. and Poole, G.J. 1981. Furrow advance rates under surge flow systems. *J. irrig. and Drainage Div.,* ASCE 107(IR3): 257-264.

Boumans, J.H., van Hoorn, J.W., Kruseman, G.P. and Tenwar, B.S. 1988. Water table control, reuse and disposal of drainage water in Haryana. Agric. Water 1VIgmt. 14: 537-545.

Craig, J.R. 1980. Saline waters: genesis and relationship to sediments and host rocks. In: Saline Water Proc. Symp. Groundwater Salinity, 46th Annual Ivleeting. R.B. Mattox (ed.). Amer. Assoc. Adv. Sci (AAAS), Las Vegas, Nevada.

Dalton, F.N. and Poss, J.A. 1990. Soil water content and salinity assessment for irrigation scheduling using time-domain reflectometry: principles and applications. Acta Horticulturae.

FAO. 1979. Yield response to water. Doorenbos, J. and Kassam, A.H. Irrigatibn and Drainage Paper 33. Rome. pp.193.

Feng, G., Zhang, Z and Zhang, Z. 2019. Evaluating the Sustainable Use of Saline Water Irrigation on Soil Water-Salt Content and Grain Yield under Subsurface Drainage Condition. *Sustainability,* 11: 2-18.

Gupta, I.C. and Pahwa, K.N. 1981. A Century of Soil Salinity Research in India - Annotated Bibliography 1863-1976. Oxford and IBH Publishing, New Delhi. pp. 400.

Gupta, I.C. 1990. Use of Saline Water in Agriculture. A study ,Df arid and semi-arid zones in India. Revised edition. Oxford and 1BH Publishing, New Delhi.

Mashali, A.M. 1989. Salinization as a major process of soil degradation in the Near East. 10th Session of the Regional Commission on Land and Water Use in the Near East, Amman, Jordan. 10-14 December 1989.

Meiri, A. and Plaut, Z., 1985. Crop production and management under saline conditions. *Plant Soil,* 89: 253-271.

Minhas, P. S. 1996. Saline water management for irrigation in India. *Agricultural Water Management,* 30: l-24.

Rawlins S.L. 1977. Uniform irrigation with a low-head bubbler system. *Agric. Water Management,* 1:167-178,

Rhoades, J. D., Kandiah, A. and Mashali, A. M. 1992. The use of saline waters for crop production. FAO Irrigation and Drainage Paper 48.

Rhoades J.D., Corwin D.L. and Hoffman G.J. 1981 Scheduling and controlling irrigations from measurements of soil electrical conductivity. Proc. ASAE Irrigation Scheduling Conf., Chicago, Illinois, 14 December 1981. pp. 106-115.

van Schilfgaarde, J. 1976. Water management and salinity. In: Prognosis of salinity and alkalinity. Soils Bulletin No. 31. FAO, Rome.

15

Remote Sensing and GIS in Diagnosis and Management of Problem Soils

Ram Gopal, D.R. Prajapati, Sunita Kumari and Vikas Rena

Remote sensing

It is the art and science of obtaining information about an object, area, or phenomenon through the analysis of data acquired by a device without being in physical contact. Remote sensing technology involves acquiring information about the earth's surface and atmosphere, using satellite based sensor, *interpretation,* of spectral measurements and *characterization* of land resources and environmental phenomena.

The information recorded by this technology is not visible to humankind. It provides synoptic view as well as three-dimensional view of larger area. And data are received periodically and helps in monitoring the changes at short intervals. The data recording is unbiased one which is stored permanently and can be used in many applications. Thus, processing of data is faster than the conventional interpretation method. Thus, soil resources data collected by extensive ground surveys and soil maps were prepared using cartographic techniques these were tedious, costly and time consuming. Multi date, multi spectral and real time satellite data provide a reliable accurate tool for assessment and repeat coverage of the same area at the same local time for monitoring soil resources (Joshi, 2013). In remote sensing, reflected energy from the object is recorded by a sensor. The information about the object is obtained through the analysis of reflected rays. Three energy interactions occurs when electromagnetic energy falls on any surface feature i.e., reflection, absorption and transmission. The interrelationship between these three energy interactions are as follows:

$$E_1(\lambda) = E_R(\lambda) + E_A(\lambda) + E_T(\lambda)$$

Where E_1 is the incident energy, E_R is the reflected energy, E_A is the absorbed energy, and E_T denotes the transmitted energy.

The proportion of energy reflected, absorbed, and transmitted will vary for different earth features, depending on their material type and condition. Secondly, the proportion of reflected, absorbed, and transmitted energy will vary at different wavelengths. Thus, two features may be indistinguishable in one spectral range and be very different in another wavelength band.

Monitoring study of natural resources are facilitated through Sunsynchronous /Polar /Natural resources satellites. These satellites moves in a low earth orbit called polar or sun synchronous orbit (800-1000 km altitude). The satellite orbit is fixed passage *Path* and cover fixed area (*Swath*). The east to west horizontal lines parallel to the equator which intersect the path are *Rows*. The ground area covered by the satellite passes is referred by specific *Path-Rownumbers*. As the satellite orbits in the north south plane the earth below spins around its axis from west to east. Therefore after completing one orbit the satellite changes its path from east to westward. The Landsat series satellite imageries have been extensively used in monitoring. The IRS satellites are mounted with Linear Imaging Self Scanning (LISS), AWiFS (Advanced Wide Field Scanners) and PAN (Panchromatic) sensors (Manikandan and Prabhu, 2011; Joshi, 2013). Sensor refer to the device that records the electromagnetic radiation reflected from the objects. The detection of electromagnetic energy can be performed either photographically or electronically. The specific parameters of sensors are spatial resolution, spectral resolution, radiometric resolution and temporal resolution.

(a) **Spectral resolution**

All objects on earth surface under natural condition reflect incident solar radiation. According to ground features, amount of spectral reflectance received in different wavelength bands and spectral reflectance is recorded by the sensor in different wavelength bands viz. visible (blue, green, red), near Infrared (NIR) and short Wave Infrared (SWIR) etc. Spectral resolution is determined by width of wavelength band. Spectral reflectance of soil increase, with increase in visible region bands and constant in near IR region band.

(b) **Spatial resolution**

The smallest area on earth surface for which the sensors can record spectral reflectance. The spectral reflectance is the smallest unit called pixel.

Spectral reflectance for different type of soil

(i) **Typical arid soils:-** Sandy, saline-sodic, shallow and gypsiferous arid soils produce light tune images so their mapping is difficult. Spectral reflectance of each of these soils with the help of Hindivac Spectro-

Radiometer-101 and BaSO4 plate as the standard reference in the wavelength range of 450-1000 nm at the interval of 50nm (Joshi, 2013).

(ii) **Salt affected soils:-** Spectral reflectance of natural salt affected soils were in this order: highest (65%) > brackish water irrigated sodic soils (37-60%)> natural saline soils (43%) >irrigated saline soils (34-51%) (Kalra and Joshi, 1994; Joshi et al., 2004; Joshi and Pagaria, 2004). Higher reflectance values were obtained at 800-1000nm wave length followed by 460 nm while lowest at 620 nm wavelength.

(iii) **Sandy soils**: -Spectral reflectance of sandy soils at wavelength 450 nm ranged between 17.5 to 25% with mean value of 21.4%. Between wavelength range 600-800nm (visible range) reflectance values showed increasing trend (mean value 22.6-28.1%) while reflectance values between 800-1000 nm (NIR) wavelength were almost stable. Spectral reflectance values of wind eroded sandy soils were higher (mean values 27.0 -27.4%) (Joshi, 2013).

Geographic Information System (GIS)

A GIS is a computer based information system for input, storage, analysis, management, manipulation, retrieval and output of geographic data and their attributes. All information in GIS must be linked to a spatial reference (Latitude/ longitude) or other spectral co-ordinates. Thus the GIS use geo-references as the primary means of storing and accessing information (Joshi, 2013).

GIS used for presenting and analyzing the geographic features present on the earth surface and the events (non-spatial attributes) that taking place on it. It proved to be an effective tool in handling spatial data available at different scales, voluminous point data such as soil information, rainfall, temperature etc. and socio-economic data and to be perform integrated analysis of data on various resources of any region and to arrive at optimum solutions for various problems (Venkataratnam and Ravisankar, 1996). Simply, it is a computer-assisted information system for the acquisition, storage, analysis and display of geographic data in a desired manner. It is designed to work with data referenced by geographical coordinates. GIS is not a single piece of software, it is made up of five key components; Hardware, Software, Data, Live ware and methodology. Now -a-days, the importance of spatial dimensions is recognized because of its organizing and analyzing capability. It has the ability to assimilate divergent sources of data, both spatial and non-spatial, and produces impacts more visually.

Problem soil is a part of wastelands

The ministry of rural development has come out with the fifth edition of Wasteland Atlas -2019. This is significant as it takes into account 12.08 Mha of unmapped area of Jammu and Kashmir (J&K) for the first time. The effort has resulted in estimating the spatial extent of wastelands for entire country to the tune of 55.76 Mha (16.96 per cent of geographical area of the country i.e. 328.72 Mha) for the year 2015-16 as compared to 56.60 Mha (17.21 per cent) in the year 2008-09. As per the Atlas, during this period 1.45 Mha of wastelands are converted into non wastelands categories. The new wastelands mapping exercise, carried out by NRSC using the Indian Remote Sensing Satellite data is the fifth edition of Wastelands Atlas-2019. India with 2.4 per cent of total land area of the world is supporting 18 per cent of the world's population. The per capita availability of agriculture land in India is 0.12 ha whereas world per capita agriculture land is 0.29 ha. According to the ministry, unprecedented pressure on the land beyond its carrying capacity is resulting into degradation of lands in the country. Therefore, robust geospatial information on wastelands assumes significance and effectively helpful in rolling back the wastelands for productive use through various land development programmes/schemes, it said. The wastelands have undergone positive change in the states of Rajasthan (0.48 Mha), Bihar (0.11 Mha), Uttar Pradesh (0.10 Mha), Andhra Pradesh (0.08 Mha), Mizoram (0.057 Mha), Madhya Pradesh (0.039 Mha), Jammu & Kashmir (0.038 Mha) and West Bengal (0.032 Mha). Majority of wastelands have been changed into categories of 'croplands' (0.64 Mha), 'forest-dense / open' (0.28 Mha), 'forest plantation' (0.029 Mha), 'plantation' (0.057 Mha) and 'industrial area' (0.035 Mha) etc. (Economic Times, By *Yogima Seth* Sharma, Nov 05, 2019).

Wastelands: It is a degraded land which can be bought under vegetation cover, with reasonable effort and which is currently under utilised and land which is deteriorating for lack of appropriate water and soil management or on account of natural causes. It is categories into two sub-divisions:

(a) Culturable Wasteland: The land which is has potential for the development of vegetative cover and is not being used due to different constraints of varying degrees, such as erosion, water logging, salinity etc.

(b) Unculturable Wasteland: The land that cannot be developed for vegetative cover for instance the barren rocky areas and snow covered glacier areas.

India with 4% of total land area of the world is supporting 18% of the world's population. The per capita availability of agriculture land in India is 0.12 ha whereas in the world per capita agriculture land is 0.29 ha. The efforts has resulted in estimating the spatial extent of wastelands for entire country to the tune of 55.76 Mha (16.96% of geographic area of the country i.e. 328.72 Mha) for the year 2015-16 as compare to 56.60 Mha (17.21%) in the year 2008-09. During this period 45Mha of wastelands are converted into non wastelands categories.

Problematic Soils: The soils which are unfavorable for cultivation of field crops because of one or more unfavorable soil properties/characteristics (viz. Soluble salts, soil reaction, ESP, water logging, aeration etc.) are adversely affect the optimum soil productivity is called problematic soils, shown in Table 15.1.

Table 15.1: The major problematic soils of India

Sl. No.	Problematic Soils	Key Diagnosis	Constraints
1.	Clay soils	Dominated by clay particles	Water logging, compaction, poor aeration, difficult to cultivate
2.	Sandy soils	Dominated by Coarse sand particles	poor fertility, low SOM, low water holding capacity, erosion
3.	Acid soils	Soil pH < 6.5	Fe, Al toxicity (Strong acid soil)
4.	Salt affected soils		
	a. Saline soils	ECe is > 4.0 dS/m, Soil pH <8.5, ESP <15, SAR<13	High osmotic potential, nutrient imbalance
	b. Sodic soils	ECe<4, Soil pH 8.5-10, ESP>15, SAR>13	Deteriorated physical condition, Na toxicity, nutrient imbalance
	c. Saline sodic soils	ECe>4.0 dS/m, Soil pH >8.5 and ESP >15, SAR>13	High osmotic potential, deteriorated physical condition, nutrient imbalance
7.	Calcareous soils	$CaCO_3$> 5.0 %	P, Fe deficiency
8.	Water logged soils	Water Stagnation, Low infiltration rate,	Poor aeration
9.	Degraded soils	Based on soil analysis	-
10.	Compacted soils	High bulk density	Poor aeration, poor root penetration, water logging
11.	Impermeable soils	Low hydraulic conductivity (HC) and infiltration rate	Poor aeration, water logging

Source: Ashok kumar V. Rajani (2019)

Remote sensing technology and GIS in India

In India, initially aerial photographs were used in deriving information on degraded lands (Kamphorstand Iyer, 1972; Iyer *et al.*, 1975). The application of remotely sensed data in mapping degraded lands fromspace borne sensors started with the launch of the first Earth Resources Technology Satellite

ERTS-l/Landsat-I. However, the satellites Landsat-TM, SPOT and Indian Remote Sensing Satellites (IRS) with better spatial and spectral resolution, enabled to map and monitor degraded lands more efficiently. Many studies are reported in literature on the use of Landsat-MSS, TM, IRS and SPOT data for inventory of degraded lands. Studies are carried out on mapping eroded lands (Venkatamtnamand Rao, 1977; Rao *et al.*, 1980, ravines (NRSA, 1981; Karale *et al.*, 1988), watershed prioritization (Dohare *et al.*, 1985, Sharada *et al.*, 1993), salt-affected soils (Venkataratnam, 1983, 1984; Bajwa *et al.*, 1990) and shifting cultivation (Kushwahaand Unni, 1987).

Elements of satellite image interpretation (Manikandan and Prabhu, 2011; Joshi, 2013)

1. Colour: The satellite images received in green, red and near Infrared bands are merged to form FCCs and assigned false color magenta-red for vegetation and blue for water bodies. Urban areas appear as light grey and hills in dark grey. Different shades of false colours help in interpretation of satellite images.
2. Size: Size give good indication of certain objects.
3. Texture: It is the frequency of arrangement of tones in a photograph. Texture may be described as rough, smooth, coarse, fine etc.
4. Tone: Refers to the distinguishable shade variation from white to black. Objects that reflect more light are observed as white tone. E.g. white objects-saline soil, sandy soil. Dark or black tone in photograph designates the object, which reflects less light. E.g. black soil.
5. Shape: Shape is one of the most important factors for recognizing the object. Hills, sand dunes, trees and irrigated fields can be identified by their shape.
6. Pattern: It relates to the spatial arrangement of the object.

Satellite image interpretation

The satellite images are available as soft copy in the form of CCTs, CD-ROM and paper prints in different bands and as False Colour Composite (FCC).

1. Visual interpretation: Hard copies/paper prints in the form of FCCs are interpreted for various features like soil types, water bodies, hill, vegetation type and land use.
2. Computer based interpretation: The soft copies of the data are interpreted by computer software using digital image processing technique. The steps involved are:

(a) Pre-processing of remote sensing data

(b) Geometric correction

(c) Image enhancement: contrast enhancement, linear contrast stretching, band rationing, Normalized Difference Vegetation Index (NDVI), Principal component analysis, filtering etc.

(d) Image classification: Supervised classification requires knowledge about ground truth. With unsupervised classification images of similar tone, texture, and pattern are classified together.

Interpretation of satellite data and ground truth: Following 24 geo-coded satellite FCC images of IRSP6LISS-III (1:50,000 scale), data acquired on May 2005 were visually interpreted in conjunction with the SOI toposheets. Physiographic units viz., hills, pediments, buriedpediment, alluvial plain viz., medium and fine textured and valley fills were delineated. Within the physiographic unit tentative soil boundaries were delineated by adopting the elements of image interpretation including tone, texture, pattern, association, and land use and vegetation pattern. Field traversing was carried for ground truth, soil augar holes were examined at close interval depending up on terrain variability, and morphological characteristics. For each soil series, at typical sites, profiles were examined and for each horizon morphological characteristics including soil colour, texture, structure, consistency and calcareousness were recorded (Soil Survey Manual, 1970). The salinity and erosion status were also observed. Soil samples of each horizon of the typical profiles were collected for analysis in the laboratory. Based on field observation and chemical analysis the soil series were recognized and classified according to Keys to Soil Taxonomy (Kalra *et al.,* 2010).

Identification of Soil Constraints through Remote Sensing and GIS

Remote sensing and GIS technology has been used for identification of soil constraints in resource potential Bhilwara district of Rajasthan. IRS LISS-III FCC images were interpreted for soil constraints using physiography soil approach, verified through field checking and laboratory analysis. On IRS LISS-III FCC images the salt affected soils of Kotri and Taswaria appeared in bright white to light grey tone,smooth texture with white mottles. These were also verified during ground truth and soil analysis for salinity (EC 2.90-3.32 dS m-1) and sodicity (pH 9.50-9.86 and ESP 17.60 -19.05). Similarly on the LISS III FCC, constraints due to water erosion near Bir, Sareriand Vijaypura soil series were apparent in light grey to whitish tone, intercepted by medium grey streaks indicating streams and exposed sub-soil. The constraints due to shallow depth associated with rockout crops and hilly areas of Balda and Delwara seriesappeared in greenish grey tone and coarse texture. Out of total geographical area of the district, 17 per cent area has constraints due to salinity/sodicity. In 48 per cent area have constraints due to uneven terrain, shallow soils mixed with gravels, low AWC, low nutrient retention, moderate-to severe erosion whichare difficult to overcome while 31 per cent area has no soil constraint, the soils have high potential, and with intensive irrigation andnutrient management high crop yield can be realized (Kalra *et al.*, 2010).

Application of GIS

The soil series maps prepared toposheet-wise for entire district at 1: 50,000 scale were scanned and boundaries for each mapping unit were digitized using ARC-INFO version 9.2 GIS software. The digitized maps have been used to generate the maps showing soil series and constraints viz., salinity/sodicity, water erosion and depth in Table 15.2. The soil series have been checked for soil/terrain constraints viz. salinity/sodicity, effective soil depth, water erosion, AWC and physical constraints as per Eswaran (1992).

Table 15.2: Assessment, monitoring and management of degraded lands

Type of Soil	Tone	Scale	Sensor
1) Salt affected soils (Venkataratnam and Ravisankar, 1996)	On the FCCs in bright to dull white tone patches within the background of normal soils supporting good vegetation (that appears as bright red/ magenta tone). The salt-affected soils with poor crop growth appear in dull red tone/mottled tone.	The salt-affected soils of India have been mapped at 1:250,000 scale through visual Interpretation. NRSA has also mapped salt-affected soils at 1:50,000 scale on a limited extent. At 1:50,000 scale, pure units could be mapped and even smallunits, not to be mapped at 1:250,000 scale, could also o be delineated.	Landsat TM/IRS imagery at the National Remote Sensing Agency (NRSA, Hyderabad) in association with the National Bureau of Soil Survey and Land Use Planning (Nagpur).
2) Saline soil (Sharma, et al., 2009	These soils appear in whitish gray, milky white, dull white and light blue tones on imagery depending upon severity and other soil constituents present.	-	IRS LISS I, II,III

Type of Soil	Tone	Scale	Sensor
3) Sodic soil (Sharma, *et al.*, 2009)	Light gray with white mottles, White to bright white mottles, Moderate white to white red mottles, chess board pattern	-	IRS LISS I, II, III
4) Waterlogged by Tawa Command area (Choubey, 1997)	Grey color	1:250,000	IRS-1A-LISS-1

GIS for management of salt-affected soils

A study has been conducted for extracting the spatial information on Salt-affected soils of Unnao district, Uttar Pradesh state using IRS-1LB1 digital data andcalculate the gypsum requirement for the reclamation of salt-affected soils for the whole district. Using GIS techniques with the help of attribute data on salt-affected soils and image analysis was carried out using EASI/PACE image analysis software package on IBM 6000 workstation. PAMAP GIS (a raster based package) was used for storing, retrieving and analysis of spatial and attribute data on a Sun work station.The satellite data were registered with the topographical maps of the district and resampled to the map through identification of sufficient number of ground control points and the district boundary was extracted. Subsequently, a mask was built for the area falling outside the district. Based on available information, training areas were defined for various land use/cover classes including salt-affected soils. After studying the spectral response of salt-affected soils and relating them with soil properties they could be grouped into strong and moderate degree classes. Then image classification was done to obtain the spatial extent of salt-affected soils in the district (Venkataratnam and Ravisankar, 1996).

Conclusion

- Spaceborne satellite data have become valuable tools in studying the spatial extent of problematic soil and for monitoring the changes that have taken place over a period of time
- Soil salinity/alkalinity significantly effects crop production and consequently has negative effects on it. Remote Sensing and GIS technology is very effective tool for suggesting action plans /management strategies for problematic soil areas.
- GIS is considered one of the important tools for decision making in problem solving with geo-information.

- In the near future GIS will become a very important tool for handling voluminous data generated on degraded lands through conventional and remote sensing techniques and for integrated analysis of data to derive plans for reclamation of problematic soil.

References

Ashok kumar V. Rajani. 2019. Problematic soils and their management (ppt. presentation) (https://www.researchgate.net/publication/335723611_Problematic_soils_and_their_management/stats).

Bajwa, S. S., Singh, C., Sharma, P. K. 1990. Monitoring of salt-affected soils of Sangrur district using remote sensing technology, Proc. Symp. Remote Sensing for Agricultural Applications, New Delhi, pp.137.

Choubey, V. K. 1997. Detection and delineation of waterlogging by remote sensing techniques. *Journal of the Indian Society of Remote Sensing,* 25(2): 123-135.

Dohare, D. D., Ali, S., Shanware, P. G. 1985. Evaluation of sedimenty yield index using Landsat data and geographic information system, Proc. Sixth ACRS, Hyderabad.

Eswaran, H. 1992. Role of soil information in meeting the challenge of sustainable land management. *Journal of the Indian Society of Soil Science*, 40: 6-24.

Iyer, H. S., Singh, A. N., Kumar, R. 1975. Problem area inventoly of parts of Hoshiarpur district through photo-interpretation, *J. Indian Photoint.,* 3(2): 79.

Joshi, D. C. 2013. Remote sensing and GIS based soil resources appraisal. *Curr. Agric.,* 37(1-2): 1-116.

Kalra, N. K., Singh, L., Kachhwah, R., Joshi, D. C. 2010. Remote Sensing and GIS in Identification of Soil Constraints for Sustainable Development in Bhilwara District, Rajasthan. *J. Indian Soc. Remote Sens.,* 38: 279–290.

Kamphorst, A. and Iyer, H .S. 1972. Application of aerial photo-interpretation to ravine surveys in India. Proc. 12th Cong. Interrlat. Soc. Photogranz. Eng. Ottawa, Canada.

Karale, R. L., Saini, K. M., Narula, K. K. 1988. Mapping and monitoring ravines using remotely sensed data India. *J. Soil Wat. Cormen.,* 32(1,2): 75.

Kushwaha, S. P. S., Madhavan, U. N. V. 1987. Impact of shifting agriculture on forest degradation and land transformation - A case study in Meghalaya using Landsat-TM data, Proc. Nat. Symp. Remote Sensing in Land Transformation and Mgmt., Hyderabad.

Manikandan, K. and Prabhu, S. 2011. Book Indian Forestry. Published by Jain Brothers (New Delhi). East Park Road, New Delhi, pp. 286-293.

NRSA, 1981. Satellite remote sensing survey for soil and land use in part of Uttar Pradesh. Project Report National Remote Sensing Agency, Hyderabad.

Rao, K. V. S., Bali, Y. P., Karale, R. L. 1980. Remote Sensing for soil conservation and watershed management, Proc. Sem. Application of Photointerpretation and Remote Sensing Techniques for Natural Resources survey and Environmental Analysis, Dehradun.

Sharada, D., Ravi, K. M. V., Venkataratnam, L. and Rao, T. C. M. 1993. Watershed prioritization for soil conservation - A GIS appproach, *Geocartao International,* (1): 27-33.

Sharma, J. R., Bothale, R. V., Bera, A. K., Bothale, V. M., Dutta, D., Paithankar, Y., Agarwal, C. K. and Singh, R. 2009. Assessment of Waterlogging and Salt and / or Alkaline affected Soils in the Commands of all Major and Medium Irrigation Projects in the Country using Satellite Remote Sensing, New Delhi. (https://www.indiawaterportal.org/ sites/indiawaterportal.org/files/introduction.pdf).

Sharma, Y. S. 2019. Economic Times, Nov. 5, 2019. (https://economictimes.indiatimes.com/news/economy/agriculture/ministry-of-rural-developmentreleases-fifth-edition-of-wasteland atlas/articleshow/71920250.cms?from=mdrhttps://www.chronicleindia.in/current-news /279-waste lands-atlas-ndash-2019)

Venkataratnam, L. 1983. Monitoring of salinity in Indo-Gangetic plains of North-Western India using multi-date Landsat data, Proc. 17th internat.Symp. Remote Sensing of Environ. ERIM, Ann Arbor, Michigan.

Venkataratnam, L. 1984. Monitoring and managing soil and land resources in India using remotely sensed data. Proc. 3rd Asian Agricultural Symp.,Chiang Mai, Thailand.

Venkataratnam, L. and Rao, K .R. 1977. Compter aided classification and mapping soils and soil limitations using Landsat multispectral data. Proc. Symp. Rernote Sensing for Hydrology, Agriculture and Mincral Resources. Space Applications Centre. Ahmedabad.

Venkataratnam, L. and Ravisankar, T. 1996. Remote sensing and GIS for assessment, monitoring and management of degraded lands. National Remote Sensing Agency, Dept. of Space, Govt. of India, Hyderabad-500037, India, pp. 503-516.

16

Multipurpose Tree Species (MPTs)

Abhay Kumar, M. S. Malik, M.H. Siddiqui and M. S. Yadava

Introduction

The term multipurpose trees (MPT) refers to all woody perennials that are grown to provide more than one significant function (shelter shade, land sustainability) of the land use system they implement. More emphasis is given to trees which are indigenous in nature and have the capability to fix atmospheric nitrogen. All trees are said to be multipurpose. The term "multipurpose" as applied to trees for agroforestry refers to their use for more than one service or production function in an agroforestry system (Burley and Wood, 1991).

Multipurpose trees species (MPT) are grown to provide more than one significant crop of function or form. These may include fuel wood, fiber, fodder, flower, fruit, fertilizer (6F) timber, industrial purpose, nitrogen fixing, soil conservation, soil protection and rehabilitation, shade and nurse trees, medicine, windbreak, shelterbelts, live fences, aesthetic or recreational purpose, shown in Table 16.1.

Selection criteria for suitable MPTs (Multipurpose trees)

- Identifying the proper species suited to the soil and site conditions
- Tree species should be fast to medium growing
- Single stemmed with clear bole and tall
- Deep rooted with tap root system so the trees do not compete with companion crops for moisture and nutrient
- Trees should have a low crown diameter
- Light branching in habit
- Little fall and litter decomposition rate should have positive effect upon the soil

- Preferably leguminous and nitrogen fixer
- Marketability of products (wood, fruit, fodder, fuel etc.)
- Resistance to disease and pests

Benefits from MPTs (multi-purpose trees)

1. Reduction of surface run-off, nutrient leaching and soil erosion
2. More efficient recycling of nutrients by deep rooted trees on the site
3. Better protection of ecological systems
4. Improvement of microclimate
5. Increment in soil nutrients through addition and decomposition of litter fall

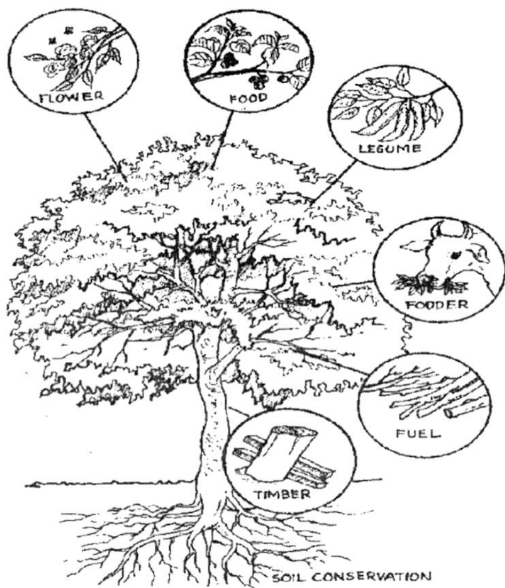

Fig. 16.1: Multipurpose tree and their uses

6. Improvement of soil structure through the constant addition of organic matter from decomposed litter
7. Increment in maintenance of outputs of food, fuel-wood, fodder, fertilizer and timber
8. Reduction in incidence of total crop failure, common to single cropping or monoculture system
9. Increase in levels of farm income.

10. Improvement in rural living standards from sustained employment and higher incomes
11. Improvement in nutrition and health due to increased quality and diversity of food outputs
12. Sustained supply of raw materials to forest based industries
13. Reduction in hunger and poverty by increasing production of farms for both consumption and sale
14. Reduction in the speed of climate change by more carbon sequestration and storage by trees

Table 16.1: Common multipurpose trees (MPTs) species in India

Sl. No.	Botanical Name	Local Name	English Name	Family	Uses
1	Acacia auriculiformis	Bengali babul	Australian Babul	Mimosaceae	Timber, fast growth, fuel wood, roadside plantation,
2	Acacia catechu	Khair	Cutch tree	Mimosaceae	Catch/Katha/gum extraction, timber, fodder, fuel wood
3	Adina cardifolia	Haldu	Hed	Rubiaceae	Timber, fodder, fuel wood
4	Aegle marmelos	Bel	Holy fruit	Rutaceae	Man edible, medicinal, religious
5	Alianthus excels	Arduso	Tree of heaven	Simaroubaceae	Timber, pulp & paper, ply wood, match strict
6	Albizia lebbeck	Black siras	Womans tongue tree	Mimoseceae	Medicinal, bark use fish poison, timber, fuel wood
7	Annona reticulate	Ramphal	Bullocks heart	Annonaceae	Man edible, fuel wood, fodder
8	Artocarpus heterophyllous	Kathal	Jack fruit	Moraceae	Man edible, timber, fuel wood, fodder, dye
9	Azadirechta indica	Neem	Margosa tree	Meliaceae	Insect repellent, antibiotic, windbreaks, mulch, timber, firewood, medicimal
10	Bambusa arundinacea	Katis vans	Thorny bomboo	Granineae	Handicraft/house/pole, making, medicinal, man edible, pickle
11	Bauhinia purpurea	Purplekanchnar	Butterfly tree	Caesalpiniaceae	Edible flower/leaves/fruit, gum, medicinal, pickle, fodder, fuel wood
12	Bombax ceiba	Lalsemal	Red silk cotton tree	Bombacaceae	Flosses, medicinal, fuel wood, match stick, pulp & paper
13	Butea monosperma	Palash	Flame of the forest	Papilionaceae	Dye, lac cultivation, timber, fuel wood
14	Cassia fistula	Amaltash	Indian laburnum	Caesalpiniaceae	Tannin, ornamental, timber, fuel wood, nitrogen fixing
15	Cassia siamea	Kashid	Yellow cassia	Caesalpiniaceae	Ornamental, timber, fuel wood, nitrogen fixing
16	Ceiba pentandra	Safedshimal	white silk cotton tree	Bombacaceae	Flosses, medicinal, fuel wood, match stick, pulp & paper

Sl. No.	Botanical Name	Local Name	English Name	Family	Uses
17	Casuarina equisetifolia	Sharu	Beef wood	Casuarinaceae	Windbreak, ornamental, tannin, furniture, timber, fuel wood, nitrogen fixing
18	Cinnamomum camphora	Kapur	Camphor tree	Lauraceae	Kapur making, medicinal, fuel wood,
19	Cinnamomum zeylanicum	Tejpatra	Cinnamomum tree	Lauraceae	Man edible, medicinal, fuel wood,
20	Citrus aurantifolia	Limbu	Lime	Rutaeceae	Edible fruit, insect repellent, antibiotic, citric acid, drinking, various market product
21	Cocos nucifera	Nariyal	Coconut	Arecaceae	Food, juice from inside the coconut, roof thatching, firewood, shade, coco pit, coir
22	Dalbergia latifolia	Shisham	Indian rosewood	Papilionaceae	Timber, fodder, fuel wood, nitrogen fixing,
23	Delonix regia	Gulmohar	Royal poinoiana	Caesalpiniaceae	Ornamental, timber, fuel wood, nitrogen fixing
24	Derris indica	Karanj	Indian beech tree	Papilionaceae	Oil extraction, medicinal, manure, fuel wood, nitrogen fixing
25	Emblica officinalis	Amla	Indian goose-berry	Euphorbiaceae	Edible fruit, vitamin-C, trifala churn, various market product
26	Eucalyptus tereticornis	Nilgiri	Eucalyptus	Myrtaceae	Essential oil, vicks, perfume, pulp & paper, ply wood, pole
27	Ficus benghalensis	Bargad	Banyan tree	Moraceae	Traditional medicine, inferior quality, rubber religious, edible fruit for wild animal
28	Ficus elastic	Indian rubber	Indian caoutechoue	Moraceae	Rubber making, traditional medicine, fuel wood
29	Ficus religiosa	Pipal	Sacred ficus	Moraceae	Traditional medicine, religious, edible fruit for bird, fodder, fuel wood
30	Gliricidia sepium	Quickstick	Gliricidia	Fabaceae	Living fences, firewood, fodder, fixing nitrogen into the soil, protein bank
31	Gmelina arborea	Gamhar, sevan	White teak	Verbenaceae	Timber, fodder, fuel wood, handicraft
32	Grewia asiatica	Phalsa	-	Malvaceae	Edible fruit, timber, fodder, fuel wood

Sl. No.	Botanical Name	Local Name	English Name	Family	Uses
33	Kigelia pinnate	Motipadad	Candle tree	Bignoniaceae	Medicinal, ornamental, timber, fodder, fuel wood
34	Lagerstromia speciasa	Taman	Pride of India	Lythraceae	Ornamental, roadside plantation, timber, fodder, fuel wood
35	Leucaena leucocephala	Subabul	Ipilipil	Mimosaceae	Living fences, firewood, fodder, fixing nitrogen into the soil, protein bank
36	Litchi chinensis	Lichi	Litchi	Sepindaceae	Edible fruit, juice, various market product, timber, fodder, fuel wood
37	Madhuca indica	Mahuda	Butter tree	Sapotaceae	Edible fruit, liqueur making, timber, fuel wood
38	Mallotus philipensis	Kumkum tree	Kamala tree	Euphorbiaceace	Dye, timber, fuel wood
39	Mangifera indica	Aam	Mango tree	Anacardiaceae	Edible fruit, juice, various market product, timber, fodder, fuel wood
40	Michelia champaca	Son champa	Golden champa	Magnoliaceae	Perfume, ornamental, timber, fuel wood
41	Moringa oleifera	Sahjan, Drumstick	House radish tree	Moringaceae	Edible leaves, pods and beans, animal forage, medicinal
42	Morus alba	Shetur	Mulberry	Moreceae	Silk production, hockey strict, fodder, edible fruit, fuel wood, pulp & paper, ply wood,
43	Murraya koenigii	Mitho	Curry leaf tree	Rutaceae	Edible leaves, animal forage, medicinal, fuel wood
44	Pinus roxburghii	Chir pine	-	Pinaceae	Resin/turpentine extraction, timber, fuel wood
45	Pithocaliabium dulce	Jangalijalebi	Manila tamarind	Mimosaceae	Timber, fodder, edible fruit, fuel wood,
46	Prosopis cineraria	Khejri, shany plant	-	Mimosaceae	Tannin/gum extraction timber, fodder, edible fruit, fuel wood, nitrogen fixing
47	Psidium guajava	Amrud	Guava	Myrtaceae	Edible fruit, juice, various market product, timber, fodder, fuel wood
48	Santalum album	Chandan	Sandal wood	Santalaceae	Perfume, essential oil extraction, medicinal and various market product

Sl. No.	Botanical Name	Local Name	English Name	Family	Uses
49	Shorea robusta	Sal	Resin tree	Dipterocarpaceae	Dammar, dona making, timber, fodder, fuel wood, religious
50	Syzygium cumini	Jamun	Black plum	Myrtaceae	Edible fruit, juice, various market product, timber, fodder, fuel wood
51	Swietenia mahogoni	Mahagoni	Mahagoni	Meliaceae	Timber, fuel wood, medicinal, musical instrument, ply wood
52	Temarindus indica	Emli	Tamarind	Caesalpiniaceae	Edible fruit, medicinal, seed oil, candy, pickle, various market product, timber, fodder, fuel wood
53	Tectona grandis	Sag	Teak	Vervenaceae	King of timber, fuel wood
54	Terminalia arjun	Arjunsader	Arjun	Combretaceae	Tannin, tasar silk production, timber, fuel wood
55	Terminalia cetappa	Deshibadam	Indian almond	Combretaceae	Edible fruit,timber, fuel wood
56	Terminalia chebula	Harde	Chebulic myrobalan	Combretaceae	Trifala churn, tannin, Edible fruit, medicinal, timber, fuel wood
57	Terminalia bellirica	Bahrda	Belleric myrobalan	Combretaceae	Trifala churn, tannin, Edible fruit, medicinal, timber, fuel wood
58	Vitex negundo	Nagod	Chwte tree	Verbenaceae	Medicinal, timber, fuel wood
59	Zizyphus mauritiana	Ber	Indian cherry	Rhamnaceae	Edible fruit, lac cultivation, medicinal, pickle, timber, fuel wood

Conclusion

All trees are multipurpose. They bring subsoil nutrients to the surface, provide shade, and slow erosion. Many trees provide fodder, living fence posts, fruit and other edible parts, shade, insecticides, and wood; they all have some role in soil stabilization and offer quality-of-life benefits like beauty and a shelter for informal gatherings. Working with trees is an important investment which can be significant to the future of your Community. This chapter gives ideas and information on the many uses of trees in agricultural systems, various species, and working with trees.

Reference

Burley, J. and Wood, P. J. 1991. A Tree for All Reasons: The Introduction and Evaluation of Multipurpose Trees for Agroforestry. ICRAF, Nairobi, Kenya.

17

Bioremediation of Polluted Soil Through MPTs

Abhay Kumar, Virendra Singh and Salil Kumar Tiwari

Introduction

Soil pollution is anything that causes contamination of soil and degrades the soil quality. It occurs when the pollutants causing the pollution decrease the quality of the soil and convert the soil inhabitable for micro and macro organisms living in the soil. Soil pollution can occur either because of human activities and/or natural processes. However, mainly it is due to human activities. The soil contamination can occur due to the presence of chemicals such as ammonia, pesticides, lead, herbicides, mercury, nitrate, naphthalene, petroleum hydrocarbons, etc. in an excess amount.

Soils can be polluted with several organic and inorganic pollutants. Organic pollutants contain hazardous persistent organic compounds viz. polychlorinated dibenzofurans (PCDFs), polycyclic aromatic hydrocarbons (PAHs), polychlorinated dibenzodioxins (PCDDs), polychlorinated biphenyls (PCBs), polychlorinated naphthalines (PCNs) and other persistent organic compounds. Inorganic pollutants contain heavy metals like lead (Pb), cadmium (Cd), mercury (Hg), zinc (Zn), copper (Cu), arsenic (As) and nickel (Ni) as well as some radioactive compounds or radionuclides. The major sources of soil pollutants are anthropogenic. It contains mainly two sources, point and diffuse; the point sources contains industrial wastes, municipal wastes, domestic wastes, medical wastes, agrochemicals, agricultural wastes, composts and sludges and nuclear wastes. Organic and inorganic soil pollutants can be toxic for plants, animals soil organisms and others. Some compound of the soil pollutants enter into the food chain and can adversely affect on human health. Soil pollutants can be also transferred to surrounding air and water through leaching, volatilization, runoff and dust storms. Quality of air and water adversely affected can be degraded. Thus, remediation of polluted soils has become an awful necessity in many areas of the world.

Sources of soil pollutants

A. **Natural-** Geogenic sources, Natural events such as volcanic eruptions or forest fires can also cause natural pollution when many toxic elements are released into the environment.

B. **Anthropogenic-**

i. Industrial activities

ii. Mining

iii. Urban and transport infrastructures

iv. Waste and sewage generation and disposal

v. Military activities and wars

vi. Agricultural and livestock activities

Causes of Soil Pollution through agricultural activities

Soil pollution can cause direct and indirect effects on human health problems, starting with skin rash, headaches, fatigue, nausea, eye irritation and potentially resulting in more serious conditions viz. neuromuscular kidney, blockage, liver damage and various forms of cancer.

i. Pesticides

ii. Chlorinated Organic toxins-like DDT

iii. Herbicides

iv. Inorganic Fertilizers

v. Livestock wastes

vi. Inferior Irrigation Practices

vii. Solid Waste

Effects of soil pollution

i. Inferior Crop Quality

ii. Acidification and crop loss

iii. Harmful Effect on Human Health

iv. Water Sources Contamination

v. Negative Impact on Ecosystem and Biodiversity

Bioremediation of polluted soils

Bio means living organisms and remediate means to solve a problem so Bio-remediation means to use biological organisms to solve an environmental problem such as contaminated soil, air and water. Bioremediation is the defined as use of biological processes to break down, degrade, transform and/ or essentially remove contaminants or impairments of quality from soil, air and water. It is a natural process which relies on bacteria, fungi, and plants to change contaminants as these organisms carry out their normal life functions.

Simply it is also a technique used to remove environmental contaminants from the ecosystem. Bioremediation utilizes the biological mechanisms inherent in micro organism and plants to eliminate hazardous pollutants and restore the ecosystem to its original form (Ayangbenro *et al.,* 2017). The fundamental principles of bioremediation involve reducing the solubility of these environmental contaminants by changing pH, organic matter, EC, the redox reactions and adsorption of contaminants from polluted environment (Jain *et al.,* 2016). Many studies have been made on enhancing bio-sorption of pentachlorophenol (PCP) by changing the pH levels in aqueous solutions. For eg., the bio-sorption capacities of *Aspergillus niger* (Mathialagan *et al.,* 2009) and *Mycobacterium chlorophenolicum* (Brandt *et al.,* 1997) in the removal of PCP from aqueous solutions were reported to be pH-dependent.

Effects of heavy metals on the environment

Heavy metals like Zn, Hg, Cr, As, Cd, Ur, Ni, Ag, Au and Se are hazardous heavy metals that contaminate the environment and adversely affect on quality of the soil, crop production, human health and other organism. These pollutants are main sources of life-threatening degenerative diseases affecting humans like Alzheimer's disease, Parkinson's disease, cancer, atherosclerosis, etc. (Muszynska *et al.,* 2015).

Mechanism of heavy metal remediation by microorganisms

Microorganisms are essential in remediation of heavy metal contaminated environments as they have a various ways to endure metal toxicity. The utilization of microorganisms to sequester, precipitate or change the oxidation state of many heavy metals, shown in Table 17.1

Bacteria produce iron chelating substances called siderophores which increase the mobility and reduces bioavailability of metals resulting in its subsequent removal from soil. Sulphate reducing bacteria like *Desulfovibrio desulfuricans* have the ability to convert sulphate to hydrogen sulphate which then reacts with heavy metals such as Cd and Zn to form insoluble forms of these metal sulphides.

Fungi are able to withstand and detoxify metal ions by active accumulation, intracellular and extracellular precipitation and valence transformation, hence they are potential biocatalysts for the bioremediation of heavy metals as they are able to absorb heavy metals into their mycelium and spores (Ayangbenro *et al.,* 2017).

Yeast (*Sacharomyces cerevisiae*) is also used as efficient agent of bioremediation because they have the ability to remediate toxic metals from polluted wastewaters by biosorption through the mechanism of ion exchange.

Algae turns out large biomass which gives them a high sorption capacity compared to other microbial biosorbents.

Table 17.1: Microorganisms used in heavy metal remediation of contaminated sites.

Class of Microorganisms	Heavy Metal Removed	References
Bacteria		
Bacillus cereus strain XMCr-6	Cr	Dong *et al.*, 2013
Pseudomonas putida	Cr	Balamurugan *et al.*, 2014
Bacillus subtilis	Cr	Balamurugan *et al.*, 2014
Kocuria flava	Cu	Coelho *et al.*, 2015
Bacillus cereus	Cr	Coelho *et al.*, 2015
Sporosarcina ginsengisoli	As	Coelho *et al.*, 2015
Pseudomonas veronii	Cd, Zn, Cu	Coelho *et al.*, 2015
Fungi		
Rhizopus oryzae (MPRO)	Cr	Sukumar *et al.*, 2010
Aspergillus fumigates	Pb	Kumar *et al.*, 2011
Aspergillus versicolor	Ni, Cu	Coelho *et al.*, 2018
Yeast		
Sacharomyces cerevisiae	Pb, Cd	Farhan *et al.*, 2015
Algae		
Spirogyra spp. *and Cladophoraspp.*	Pb, Cu	Coelho *et al.*, 2015
Spirogyra spp. *and Spirullina spp.*	Cr, Cu, Fe, Mn, Zn	Coelho *et al.*, 2015
Hydrodictylon, Oedogonium and Rhizoclonium spp.	As	Srivastava *et al.*, 2015

Polluted Soils Remediation by Multipurpose Tree Species (MPTs)

The planting of trees that have good resistance to high levels of toxic substances and a high capacity to collect and store pollutants can also be a good practice for bioremediation process in soils (Paz-Alberto and Sigua, 2013). According to Wislocka *et al.*, 2006, the most popular trees exhibiting a high capacity to accumulate heavy metals are silver birch (*Betula pendula*), alder (*Alnus tenuifolia*), black locust (*Robinia pseudoacacia*), willow (*Salix sp.),* and conifer trees, shown in Table 17.2. *Atriplex halimus* and *Phagnalon saxatile* presented phytotoxic concentrations of Zn in leaves. Therefore, the use of these species should be avoided in soils contaminated with high concentrations of Zn.

Phytoremediation deals with the cleanup of organic pollutants and heavy metal contaminants using plants and rhizospheric microorganisms. It is inexpensive, eco-friendly and an efficient means of restoration of polluted environments especially those that of heavy metals. Plants utilized in phytoremediation are the hyper accumulators with a very high heavy metal accumulation potential and little biomass efficiency. Several processes are used to remove heavy metals from contaminted soils by some plants. Phytoextraction involves the uptake and movement of metal pollutants in the soil through plant roots into above-ground components of the plants, based on the mechanism of hyperaccumulation. Based on these criteria, about 500 taxa have been identified as hyperaccumulators of some metals and the popular ones are representatives of the following families: *Brassicaceae, Caryophylaceae, Violaceae, Fabaceae, Euphorbiaceae, Lamiaceae, Asteraceae, Cyperaceae, Poaceae, Cunouniaceae and Flacourtiaceae* (Muszynska *et al.,* 2015)

Table 17.2: Multipurpose Tree species (MPTs) used in pollute soils remediation of contaminated sites.

Heavy metal/pollutant removed	Multipurpose Tree species (MPTs)	References
Organic contaminants (mineral oil and PAHs) and heavy metals (Cd, Cu, Pb and Zn)	*Salix viminalis*	Vervaeke *et al.,* 2003
Cd and Zn	*Salix, Populus* and *Alnus*	Christopher *et al.,* 2006
Total petroleum hydrocarbons (TPH)	*Acacia angustissima, Acacia auriculiformis, Acacia holosericea, Acacia mangium, Mimosa artemisiana, Mimosa caesalpiniifolia* and *Samanea saman*	Bento *et al.,* 2012.
Pb	*Mimosa caesalpiniaefolia, Erythrina speciosa* and *Schizolobium parahyba*	de Souza *et al.,* 2012
Cr	*Dendrocalamus strictus* (34.44 μg g/dw)	Sinha *et al.,* 2013
Reduction in Zn, Ni, Pb, and Cu was 41%, 48.39%, 61.60%, and 52.72%, respectively, in the polluted soil	*Leucaena leucocephala*	Adanikin *et al.,* 2018

Heavy metal/pollutant removed	Multipurpose Tree species (MPTs)	References
Organic contaminants	*Populus species* and *Myriophyllium spicatum*	Paulo *et al.*, 2018
Se	*Astragalus bisulcatus* and *Stanleya pinnata*	
Hg	*Arabidopsis thaliana, Nicotiana tabacum, Liriodendron tulipifera* and *Brassica napus*	
Cu, Ni, Zn/Cd and As, respectively	*Elsholtzia splendens, Alyssum bertolonii, Thlaspi caerulescens* and *Pteris vittata*	
Heavy metals or radioactive elements mine area As, Sb, Cu, Fe, Pb, Zn, Ni and W	*Salix, Populus, Lemna, Callitriche, Pinus pinaster, Digitalis purpurea, Quercus suber* and *Eucalyptus globulous*	
Zn and Cd	Salicaceae family members with up to 950 mg Zn/kg dry weight (DW) and 44 mg Cd /kg DW in leaves of *Populus tremula* × *Populus tremuloides.*	Migeon *et al.*, 2019
As, Cd, Cr, Cu, Hg, Ni, Pb, V, and Zn	*Amaranthus spp., Ricinus communis, Typha domingensis, Nerium oleander* and *Phragmites australis*	Al-Thani *et al.*, 2019
1,4Dioxane (dioxane)	Hybrid poplar (*Populus deltoides* × *nigra*)	Eric *et al.*, 2020

Out of the total geographical area of 329 million ha of the country, it is estimated that only 266 m ha possess any potential for production. Of this, 143 m ha is agricultural land and it is estimated that 85 m ha suffers from varying degrees of soil degradation. Of the remaining 123 m ha, 40 m ha are completely unproductive and 83 m ha is classified as forest land. Thus out of 266 m ha, about 175 m ha or 66% is degraded to varying degrees.

The following approaches are considered necessary for the land improvement (Table 17.3)

i. Agroforestry (MPTs)
ii. Watershed management
iii. Soil conservation
iv. Integrated management of natural resources
v. Improving agricultural production
vi. Promotion of sustainable land management techniques to climate change adaptation
vii. Adding value to agricultural, forestry, farmed products

Table 17.3: Multipurpose Tree species (MPTs) suitable for different westeland /barranland / polluted soils.

Type of soils	Multipurpose Tree species (MPTs)
Sandy soil area	*Casurina equisetifolia, Anacardium occidentale, Eucalyptus spp. Dalbergia sisoo, Melia azadiracta Acacia auriculiformis etc.*
Acidic soil area	*Xylia xylocarba, Alianthus ultisima, Gmelina arborea, Gliricedia sepium, Albizia procera etc.*
Alkaline soil area	*Prosopis juliflora, Luecaena leucocephala Salvadora persica, Tamarindus indica, Tamarix articulate, Acacia nilotica, Alianthus excelsa, Terminalia arjuna etc.*
Saline soil area	*Prosopis juliflora, Luecaena leucocephala Salvadora persica, Tamarindus indica, Tamarix articulate, Acacia catechu, Acacia nilotica, Alianthus excelsa, Terminalia arjuna etc.*
Laterite soil area	*Tectona grandis, Shorea robusta, Dendrocalamus strictus, Santalum album, Xylia xylocarba, Adina cordifolia, Anacardium occidentale, Eucalyptus spp. etc.*
Mining area	*Pinus caribaea, Acacia auriculiformis, Albizia lebbeck, Albizia chinensis, Albizia procera, Cassia siamea, Grevillia robusta, Dalbergia sisoo, Sterculia urence etc.*
Waterlogged soil area	*Syzygium cumuni, Terminalia arjuna, Salix tetrasperma, Acacia nilotica, Anthocephalus cadamba, Eucalyptus robusta, Dalbergia latifolia, Eucalyptus camaldulensis, Eucalyptus grandis etc.*
Dry land area	*Dalbergia sisoo, Dendrocalamus strictus, Eucalyptus territicornis, Melia azadiracta, Cassia siamea, Hardwickea binata, Acacia nilotica, Albizia lebbeck, Zizyphus mauritiana, Prosopis juliflora etc*
Rocky land area	*Prosopis juliflora, Sterculia urence, Cactus spp. etc.*
Gullied and ravine land area	*Syzygium cumuni, Terminalia arjuna, Salix tetrasperma, Dalbergia sisoo, Morus alba, Dendrocalamus strictus etc.*

Conclusion

The effects of heavy metals contamination caused by some human activities on the environment, the possible health hazards, as well as the various mechanisms and enzymatic reactions used by plants and microbes to effectively remediate polluted environments. It revealed the usefulness of bioremediation as a better substitute for the removal of heavy metals from contaminated sites compared to the physico-chemical methods which are less efficient and expensive due to the amount of energy expended. Microorganisms and multipurpose tree species (MPTs) possess inherent biological mechanisms that enable them to survive under heavy metal stress and remove the metals from the environment. Various publications found that *Salix spp.* and *Populus* spp are used for bioremediation of polluted soils in most of countries.

References

Adanikin B., Kayode J. 2018. Bioremediation Effects of Nitrogen Fixing Trees on Nutrients and Heavy Metals in Spent Engine Oil Polluted Soil. In: Leal Filho W. (eds) Handbook of Climate Change Resilience. Springer, Cham. pp. 1-9.

Al-Thani, R. F. and Yasseen, B. T. 2019. Phytoremediation of polluted soils and waters by native Qatari plants: Future perspectives, Environmental Pollution. P. 259.

Ayangbenro, A. S. and Babalola, O. O. 2017. A new strategy for heavy metal polluted environments: A review of microbial biosorbents. *Int. J. Environ. Res. Public Health,* 14: 94. doi: 10.3390/ijerph14010094.

Balamurugan, D., Udayasooriyan, C. and Kamaladevi, B. 2014. Chromium (VI) reduction by Pseudomonas putida and Bacillus subtilis isolated from contaminated soils. *Int. J. Environ. Sci.* 5: 522.

Bento *et al.,* 2012. Selection of Leguminous Trees Associated with Symbiont Microorganisms for Phytoremediation of Petroleum-Contaminated Soil. *Water Air Soil Pollut,* 223:5659–5671.

Brandt, S., Zeng, A. P. and Deckwer W .D. 1997. Adsorption and desorption of pentachlorophenol on cells of mycobacterium chlorophenolicum PCP-1. *Biotechnol. Bioeng.*, 55: 480-489.

Christopher, J. F., Dickinson, N. M. and Putwain, P. D. 2006. Woody biomass phytoremediation of contaminated brownfield land. *Environmental Pollution,* 141(3): 387-395.

Coelho, L. M., Rezende, H. C., Coelho, L. M., de Sousa, P. A., Melo, D. F. And Coelho, N. M. 2015. Bioremediation of polluted waters using microorganisms. In: Shiomi N., editor. Advances in Bioremediation of Wastewater and Polluted Soil. InTech; Shanghai, China.

de Souza, S. C. R., de Andrade, S. A. L., de Souza, L. A. and Schiavinato, M. A. 2012. Lead tolerance and phytoremediation potential of Brazilian leguminous tree species at the seedling stage, *Journal of Environmental Management*, 110: 299-307.

Dong, G., Wang, Y., Gong, L., Wang, M., Wang, H., He, N., Zheng, Y. And Li, Q. 2013. Formation of soluble Cr (III) end-products and nanoparticles during Cr (VI) reduction by bacillus cereus strain XMCr-6. *Biochem. Eng. J.*, 70: 166–172. doi: 10.1016/j.bej.2012. 11.002.

Eric, W., Aitchison Sara, L., Kelley Pedro, J. J., Alvarez Jerald, L. and Schnoor. 2020. Phytoremediation of 1,4-Dioxane by Hybrid Poplar Trees. *Water Environment Research,* 72(3):

Farhan, S. N. and Khadom, A. A. 2015. Biosorption of heavy metals from aqueous solutions by *Saccharomyces cerevisiae. Int. J. Ind. Chem.,* 6:119–130. doi: 10.1007/s40090-015-0038-8.

Jain, S. and Arnepalli, D. 2016. Biominerlisation as a remediation technique: A critical review; Proceedings of the Indian Geotechnical Conference (IGC2016); Chennai, India.

Kumar, R. R., Congeevaram, S. and Thamaraiselvi, K. 2011. Evaluation of isolated fungal strain from e-waste recycling facility for effective sorption of toxic heavy metal Pb (II) ions and fungal protein molecular characterization-A mycoremediation approach. *Asian J. Exp. Biol. Sci.* 2: 342-347.

Mathialagan, T. and Viraraghavan T. 2009. Biosorption of pentachlorophenol from aqueous solutions by a fungal biomass. *Bioresour. Technol.,* 100: 549-558. doi: 10.1016/j. biortech.2008. 06.054.

Migeon, A., Richaud, P., Guinet, F., Chalot, M. and Blaudez, D. 2009. Metal Accumulation by Woody Species on Contaminated Sites in the North of France. *Water Air Soil Pollut,* 204: 89–101.

Muszynska, E., and Hanus-Fajerska, E. 2015. Why are heavy metal hyperaccumulating plants so amazing? Biotechnol. *J. Biotechnol. Comput. Biol. Bionanotechnol.*, 96: 265-271. doi: 10.5114/bta.2015.57730.

Ojuederie, O. B. and Babalola, O. O. 2017. Microbial and Plant-Assisted Bioremediation of Heavy Metal Polluted Environments: A Review. *Int J Environ Res Public Health*. 14(12): 1-33.

Osman K.T. 2018. Polluted Soils. In: Management of Soil Problems. Springer, Cham, pp. 333-408.

Paulo, J. C. Favas, J. P., Varun, M., D'Souza, R. and Paul, M. S. 2014. Phytoremediation of Soils Contaminated with Metals and Metalloids at Mining Areas: Potential of Native Flora, Environmental Risk Assessment of Soil Contamination, Maria C. Hernandez-Soriano, Intech Open, DOI: 10.5772/57469.

Paz-Alberto, A. M. and Sigua, G. C. 2013. Phytoremediation: A Green Technology to Remove Environmental Pollutants. *American Journal of Climate Change,* 02(01): 71-86. https://doi.org/10.4236/ajcc.2013.2100

Sinha, S.,. Mishra, R. K., Sinam, G., Mallick, S. and Gupta, A. K. 2013. Comparative Evaluation of Metal Phytoremediation Potential of Trees, Grasses, and Flowering Plants from Tannery-Wastewater-Contaminated Soil in Relation with Physicochemical Properties. *Soil and Sediment Contamination: An International Journal*, 22(8): 958-983.

Srivastava, S. and Dwivedi, A. K. 2015. Biological wastes the tool for biosorption of arsenic. *J. Bioreme and Biodegrad,* 7: 2. doi: 10.4172/2155-6199.1000323.

Sukumar, M. 2010. Reduction of hexavalent chromium by rhizopus oryzae. *Afr. J. Environ. Sci. Technol,* 4:412–418.

Vervaeke, P., Luyssaert, S., Mertens, J., Meers, E., Tack, F. M. G. and Lust, N. 2003. Phytoremediation prospects of willow stands on contaminated sediment: a field trial. *Environmental Pollution,* 126(2): 278-282.

Wislocka, M., Krawczyk, J., Klink, A. and Morrison, L. 2006. Bioaccumulation of Heavy Metals by Selected Plant Species from Uranium Mining Dumps in the Sudety Mts., Poland. *Polish Journal of Environmental Studies,* 15(5): 811–818.

18

Land Capability and Sustainability Classification

Bijay Kumar Singh and P.R. Oraon

Introduction

Land capability classification is one of the very important issues in terms of sustainable land use. Land is the most valuable natural resource which needs to be harnessed according to its potential. Land use planning and management is a known strategy for achieving sustainable development. A properly prepared land use plan based on sound scientific and technical procedures, and land utilization strategies can summarize rationally the future demands. This accompanied by a strong planning process, where communities and various stakeholders are involved, can strengthen the decision making process on allocation and utilization of land resources. Land capability is referred as suitability of the land based on inherent characteristics of the soil associated land features, and climatic conditions that limit their safe use under agriculture, forestry etc. Land capability classification is a system of grouping soils primarily on the basis of their capability to produce common cultivated crops and pasture plants without deteriorating over a long period of time. The capability classification is one of a number of interpretive groupings of soils made primarily for agricultural purposes, mainly based on the inherent soil characteristics, external land features and climatic factors that limit the use of land for agricultural purposes. The land capability is governed by the different land attributes such as the type of soil, its depth and texture, underlying geology, topography, hydrology, etc. It also includes soil irrigation classes and ground water table position to decide overall suitability of lands for sustained use under irrigation. The classification points out automatically the possibilities and limitations of the climate and soil for each crop and type of agriculture.

As with all interpretive groupings, the capability classification begins with the individual soil-mapping units, which are building stones of the system. Land resource inventory is mandatory tool to decide soil health and crop productivity. There is a long history of establishing what soils are capable of (FAO, 1976), which is often described as land suitability. More recently the new concept of Soil Security, proposed by McBratney *et al.* (2013), clearly states that the soil's suitability needs to be described using the two dimensions identified as the soil's capability and its condition. These two dimensions recognize that soil is expected to perform up to seven functions, of which its ability to produce biomass and to store and supply water and nutrients are relevant here. Primarily, the capability is concerned with the soil properties that change over longer periods of time and are often not easily or economically manageable. While assessing the soil's condition is contemporary and involves measure of soil properties that change over short time periods and are manageable.

Soil as a natural resource is the nation's strength that satisfies the human needs at a given time and space. From time immemorial, water is present on the earth, which appears as a curse in the form of flood, but as a resource when used for drinking or irrigation or fishery or even a source of energy in hydro-power generation. Similarly, soil is the most powerful natural resource that contributes to survival and nourishment of all living beings on the earth. However, it is very dynamic in its stage and development due to interactions with surrounding environment (soil site), upwelling and receding water, incoming radiation, diurnal changes in weather components and biodiversity at both macro and micro levels. Soil cannot be a waste and it does not need any rest, though its existence is at risk, if it is degraded or even eroded. The mode of weathering of type specific rocks and minerals leading to a given soil formation depends not only on parent material, relief, biosphere, climate and time, but also on solar radiation as well as management inputs applied Mishra (Bhattacharyya *et al.* 2009).

Agroforestry systems are generally perceived to be sustainable and to enhance soil properties. Growing trees in conjunction with annual crops or pastures is believed to provide a more thorough plant cover to protect the soil from erosion and a deeper or more prolific root system to enhance nutrient cycling. Shortly after its creation, ICRAF organized a state-of-the-art review of soils research and agroforestry (Mongi and Huxley, 1979). A dearth of available, solid research data about the impact of agroforestry on soils and vice versa was obvious at that time. Since then, soil research in the tropics has gained considerable momentum, as has soils research in agroforestry (Nair, 1984, 1987; Young, 1986 a, b, c, 1987; Young *et al,* 1986).

The changes and variations in the natural resources should be monitored both at regional and local scale. The future of mankind is dependent on planned development and simultaneously not exploiting the natural resources like soil and water. They play a major role in governing the livelihood (Panhalkar *et al.,* 2012). India, which is agriculture dominated country having large rural population, the drawbacks and constraints, which hinder the sustainable usage of resources, is due to the unavailability of large spatial databases and inadequate monitoring techniques. The increase in the unplanned urban expansions, industries, agriculture expansion and destruction of forests has contributed to the soil erosion (Biard and Baret, 1997).

Concept of land capability and suitability

The most important aspects of land are its role in providing anchorage space to all resources, and the fact that most human activities take place on land. Land is limited in supply and there is competition for its use. In the light of increasing population, the demand is also increasing, thus optimum use of land has become a necessity. Land is also unevenly distributed in terms of its qualities. It has its limitations for different uses. Land can be improved for a particular use by certain measures, it can also be improved by a certain kind of land-use or at least sustained production can be assured. Land can deteriorate by its mismanagement, wrong land-use or by certain cultivation practices, thus in order to avoid misuse and wastage of land. It should be used judiciously considering its capability and suitability for particular use.

The importance of land capability and land suitability studies lies in their application for the purpose of agricultural land-use planning. The planning could be done broadly for large areas using classes and sub-classes or for smaller areas using units as categories. Sharma (1972) defines, "land capability classification as a field investigation of soil properties, slope, degree of soil erosion and changing Land-use patterns which form the basis for future planning of soil and water conservation". According to Siddiqi (1971) "Land classification attempts at integrating lands according to their native characteristics, pre-existing use, yield capacity, and those classes envisaged in the plan".

Land capability classification relates to climatic factors, soil characteristics, slope, and degree of erosion, water supply and drainage and similar physical environmental conditions to the Land-use and productivity. The Soil classification during ancient period in India was based on whether soil is fertile or sterile i.e. *Urvara* and *Usara*. As a matter of fact, India is diversified in geology, relief, climate and vegetation that yield a variety of soil groups. The criteria being applied to classify the Indian soils are based on geology, fertility, relief, chemical composition and physical structure, but with inherent drawback. Even in the district gazetteers and official records during British rule in India, land revenue collection was based on assessment of inherent soil fertility alone without attempting to classify the soil types of India. Prior to independence, limited effort was made to present a systematic, accurate, complete, comprehensive, and need based database for grouping of Indian soils, though Vishwanath and Ukil, 1943 made commendable efforts to classify the Indian soils.

The first Land Capability Classification (LCC) was developed by the Soil Conservation Service (now called the Natural Resource Conservation Service) in the late 1930s and early 1940s. In time, it became the agency's main tool for evaluating appropriate uses of farmland and making recommendations on soil conservation practices, although it is only one of many possible interpretations that are made from a soil survey. The LCC was a three level classification consisting of Capability Class, Capability Subclass, and Capability Unit. The system classified land into eight classes, designated by Roman numerals I through VIII, with increasing limitations on land use and the need for conservation measures and careful management. The objective of the classification is to recognize the land into a unit with similar kinds and degree of limitations and potentials. Land was placed in a Class based on landscape, slope of the field, and soil depth, texture and acidity. Only the first four classes of the LCC are considered as suitable for cropland, the remaining four classes, V through VIII, were suitable for pasture, range, woodland, wildlife, recreation, and aesthetic purposes, shown in Table 18.1.

Land Capability Assessment:- The system consist of eight classes which classes land on the basis of an increasing soil erosion hazard and decreasing versatility of use.

Class I-land is the best and the most easily farmed land and has no hazard or limitation for use, while in class VIII land nothing of economic value can be produced, and it may need protection and management to conserve other more valuable lands and watersheds.

Class I- land is the best and the most easily farmed. It has few limitations that restrict its use.

Class II- land has moderate limitations that reduce the choice of crops. It needs simple soil and water conservation practices and requires some attention to soil management.

Class III- land has severe limitations for use, hence it needs intense soil and water conservation treatment and requires careful soil management. Graded terraces are made on moderate slopes.

Class IV- land has very severe limitations. The soil and water conservation practices are more difficult to apply and maintain.

Class V- land has all the characteristics of class I land except for limitations of water-less and stoniness or rockiness or adverse climatic conditions which make it unsuitable for cultivation of crops. However, for grazing, pasture development and forestry, there are no limitations for use.

Class VI- land has the same limitations as class IV land except that they are more severe and the land is steeper.

Class VII- land has severe limitations for grazing and forestry. The land is very steep and very severely eroded, cut up into gullies and is either too wet or too dry. The land is best utilized under forest and permanent vegetation and for limited grazing.

Class VIII- lands are very steep or rough or stony or barren. To determine the land capability classes, soil texture, soil depth, drainage, coarse fragments, erosion hazards and slope are taken in to consideration for study area.

Table 18.1: Land Capability Classes

Land Capability Classes		
Land Class	Land Suitability	Land Definition
Class I	Regular Cultivation	No erosion control requirements
Class II	Regular Cultivation	Simple requirements such as crop rotation and minor strategic works
Class III	Regular Cultivation	Intensive soil conservation measures required such as contour banks and waterways
Class IV	Grazing, occasional cultivation	Simple practices such as stock control and fertilizer application
Class V	Grazing, occasional cultivation	Intensive soil conservation measures required such as contour ripping and banks
Class VI	Grazing only	Managed to ensure ground cover is maintained
Class VII	Unsuitable for rural production	Green timber maintained to control erosion
Class VIII	Unsuitable for rural production	Should not be cleared, logged or grazed

Source: Soil Conservation Services of NSW, 1988.

Table 18.2: Land Use pattern

Symbol	Description
CC	Crop cultivation
GT	Grass land with trees
HP	Horti-pasture
AF	Agro forestry (afforestation)
C1R	Single crop rainfed
C2R	Double crop rainfed
C1I	Single crop irrigated
C2I	Double crop irrigated
W1	Wasteland (cultivable)
W2	Wasteland (uncultivable)

Source: Conservation Measures, 2008.

Subclasses were identified for special limitations such as (e) erosion, (w) excess wetness, (s) problems in the rooting zone, and (c) climatic limitations. At the lowest level, Land Capability Units were identified as groupings of soils with similar levels of yield and common requirements for land management.

Procedures to classify soils according to the LCC first involved making a detailed soil survey, with additional information on slope, erosion, and land use. This information was then translated into the Land Capability Classes, with Subclasses to show particular limitations and problems, and Units to provide interpretive information for the farmer. These interpretations were often done by multidisciplinary teams consisting of agronomists, biologists, economists, engineers, foresters, range experts, soil scientists, and soil conservationists. Recommendations for farmers were often taken from standardized Capability tables and guides who were made available to all field offices. The LCC has been eclipsed by other methods of interpreting soil surveys for planning conservation practices, although these systems are still used. The system was found lacking for forestry and rangeland management, and experts in these areas developed woodland site classifications and range land management plans.

Standardization of the parameters

The land capability classification was carried out and customized on the basis of depth, texture, erosion, slope, and permeability. Therefore, it was necessary to prepare the algorithm for each of these parameters so that we can allocate the capability classes. Land allocation in a particular capability class was done on the basis of standardization of the above parameters, which showed the kind of limitations that can be accepted for a particular class. Thus, the standardization of the depth, texture, erosion, slope and permeability is given as follows so that we can understand for a particular parameter which class of the land capability has to be given. The standardization was further also helpful in preparing the weightages for these factors.

Texture of the soil

The soil texture indicates the proportion of sand, silt and clay content in the soil. For classification point of view, this feature is grouped in the classes given in Table 18.3.

Table 18.3: The categories of the texture of the soil

Class	Description	Soils
A	Very heavy	Heavy clay, 60 % or more clay particle.
B	Heavy	Clay, silty clay, sandy clay
C	Moderately heavy	Silt clay loam, clay loam, sandy clay loam.
D	Medium	Silt loam, loam, very fine sandy loam.
E	Moderately light	Sandy loam, fine sandy loam.
F	Light	Loamy fine sand, loamy sand.
G	Very light	Sand, coarse sand

Source: Laxman Rao, 2001.

Standardization of texture parameters for land capability classification is given in Table 18.4. Various texture classes and which class of land capability has to be assigned to those classes are also given.

Table 18.4: Standardization of texture parameter for land capability classification

SI. No.	Description	Data file entry	LCAP weightages
1.	Silty loam	Sil	I, II, V
2.	Loam	L	I, v
3.	Clay loam	Cl	I, II, V
4.	Silt	Si	I, v
5.	Sandy loam	SI	I, II, V
6.	Sandy clay loam	SCI	I, II, V
7.	Silty clay loam	SiCl	II
8.	Sandy clay	SC	III
9.	Silty clay	SiC	III
10.	Clay	C	III
11.	Loamy sand	LS	III
12.	Gravel	G	IV
13.	Sandy	S	IV, VI, VII

Source: Laxman Rao, 2001.

Erosion

For grouping the soil in different capability classes, the erosion is counted in terms of depth/quantity of soil removed from the land surface. It is divided into four classes for categorizing the land as given in Table 18.5.

Table 18.5: Classes for categorizing of land on the basis erosion

Class	Erosion phase	Characteristics
el	Not apparent or slight (sheet erosion)	0 – 25% top soil or original plough layer with-in a horizon removed
e2	Moderate (sheet and rill)	25 –75% top soil removed
e3	Severe (sheet, rill and small gullies)	75 – 100% top soil and up to 25% sub soil removed
e4	Very severe (shallow gullies)	Gullied land
e5	Very, very severe (shallow gullies)	Very severely gullied land or sand dunes

Source: Conservation Measures, 2008.

Standardization of erosion parameters for land capability classification is given in Table 18.6. These land capability weightages were used in this study for generating the land capability map.

Table 18.6: Standardization of erosion parameter for land capability classification

Sl.No.	Description	Data file entry	LCAP weightages
1	No to slight	No to slight	I
2	Moderate	Moderate	II
3	Severe	Severe	IV
4	Very severe	Very severe	VI, VII
5	Moderate to severe	Moderate to severe	III
6	Severe to very severe	Severe to very severe	V

Source: Laxman Rao, 2001.

Slope

It refers to the general slope of the land, divided into the categories given in Table 18.7.

Table 18.7: Classes for categorizing of land on the basis slope

Slope class	Slope range (%)	Description
1	0 to 1	Nearly level.
2	1 to 3	Very gently sloping.
3	3 to 5	Gently sloping
4	5 to 10	Moderately sloping
5	10 to 15	Strongly sloping
6	15 to 25	Moderately steep to steep
7	25 to 33	Steep
8	33 to 50	Very steep
9	More than 50	Very very steep

Source: Laxman Rao, 2001.

Standardization of slope parameters for land capability classification is given in Table 18.8. Various slope classes and which class of land capability has to be assigned to those classes are also given.

Table 18.8: Standardization of slope parameter for land capability classification

SI.No.	Description	Data file entry	LCAP weightages
1	Nearly level	Nearly level	I, I
2	Very gentle	Very gentle	III
3	Gentle	Gentle	IV
4	Moderate	Moderate	V,VI
5	Strong	Strong	VII

Source: FAO, 1976. A framework for land evaluation. Soils Bulletin, 32. FAO, Rome.

Permeability of the sub-soil

The permeability of the soil is expressed as the rate of water flow as "mm per hr" through undisturbed saturated soil, under a water head of 12.5 mm. The permeability of sub-soils divided into seven groups as given in Table 18.9.

Table 18.9: Classes for categorizing of land on the basis permeability

Description (class)	Rate of flow (mm/hour)
Very slow	Less than 1 .25
Slow	1.25 to 5
Moderately slow	5 to 20
Moderate	20 to 65
Moderately rapid	65 to 125
Rapid	125 to 250
Very rapid	More than 250

Source: Laxman Rao, 2001.

Standardization of permeability parameters for land capability classification is given in Table 18.10 Various permeability classes and which class of land capability has to be assigned to those classes are also given.

Table 18.10: Standardization of permeability parameter for land capability classification

SI.No.	Description	Data file entry	LCAP weightages
1	Moderate to rapid	Moderate to rapid	I
2	Moderately slow to moderately rapid	Moderately slow to moderately rapid	II
3	Slow to rapid	Slow to rapid	III
4	Very slow to very rapid	Very slow to very rapid	IV
5	Slight	Slight	VI, VII

Depth

Different classes of depths with their data file entry and for which depth class which value of land capability has to be assigned in Table 18.11.

Table 18.11: Standardization of depth parameter for land capability classification

SI.No.	Description	Data file entry	LCAP weightages
1.	Very deep	Very deep	I
2.	Deep	Deep	II
3.	Moderately deep	Moderately deep	III
4.	Shallow	Shallow	IV, V
5.	Very shallow	Very shallow	VI, VII

Source: Laxman Rao, 2001.

Soil classification map of Jharkhand

Agro-climatic Zones

The state has three agro climatic sub zones viz. Central and North Eastern Plateau Sub-Zone (Zone-IV), Western Plateau Sub-zone (Zone-V) and South Eastern Plateau Sub-Zone (Zone-VI). All these fall under agro-climatic zone 7 in Table 12.

Table 18.12: Distribution of different districts in three climatic sub zones

Name of the Zone	Name of the Districts
Sub Zone IV	Dumka, Deoghar, Godda, Sahebganj, Pakur, Hazaribagh, Koderma, Jamtara, Chatra, Giridih, Dhanbad, Bokaro, and 2/3rd of Ranchi
Sub Zone V	Palamu, Latehar, Lohardaga, Garhwa, Gumla, Simdega and 1/3rd of Ranchi
Sub Zone VI	East Singhbhum, West Singhbhum and Saraikela Kharsawan

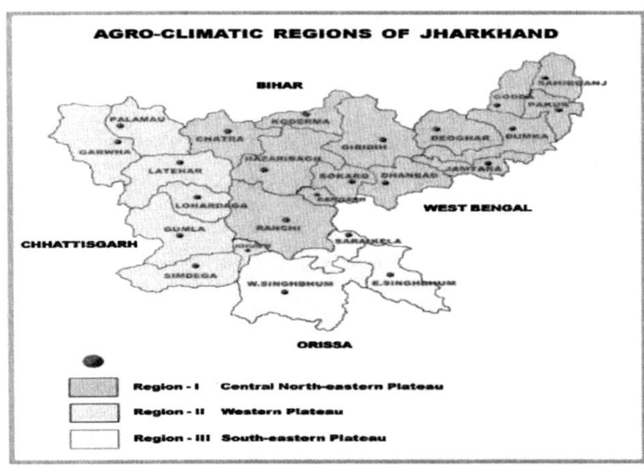

Fig. 18.1: Agro-climatic Regions of Jharkhand

Based on soil classification soil drainage map was prepared as shown in Table 18.13. Soil classes were again reclassified based on their erosion property and the corresponding map. Table 18.14 shows the area statistics of individual erosion type. Soil drainage was divided into 5 classes depending upon the drainage properties of soil and then their area of occurrence found out. It was observed that area occupied by excessively and well-drained soil is 49,996 km^2 whereas poorly drained and imperfectly drained soil 14,460 km^2 indicating that majority of the soil does not allow infiltration of water and helps in quick runoff. Severe erosion has been found in 22,639 km^2 of study area whereas moderate erosion dominates with 46857 km^2 of area. When soil erosion and drainage characteristics was clubbed up and grouped together to get the drainage stress class. Table 18.15 shows the stress map and area statistics based on soil drainage and erosion.

Table 18.13: Soil Drainage Characteristics of Jharkhand

Soil Drainage Characteristics of Jharkhand		
S.No.	Drainage Class	Area in km^2
1.	Excessively Drained	9676
2.	Well Drained	40320
3.	Moderately Drained	14854
4.	Poorly Drained	7363
5.	Imperfectly Drained	7097
	Grand Total	79310

Hazra *et al., 2011.*

Table 18.14: Erosion Characteristics of Jharkhand

Erosion Characteristics of Jharkhand		
S.No	Erosion Characteristic Area in	km^2
1.	Slight Erosion	9814
2.	Moderate Erosion	46857
3.	Severe Erosion	22639
	Grand Total	79310

Hazra *et al.* 2011

Table 18.15: Drainage Stress Class of Jharkhand

Drainage Stress Class	
Drainage Stress Area in	km^2
Excessively Drained + Moderate Erosion	1665.2
Excessively drained + Severe Erosion	8010.7
Imperfectly drained +moderate erosion	6840.2
Imperfectly drained + slight erosion	256.6
Moderately drained + Moderate Erosion	12211.1
Moderately drained + severe erosion	1773
Moderately drained + slight erosion	870.5
Poorly drained + Moderate Erosion	2571.3
Poorly drained + slight erosion	4791.4
Well drained + Moderate Erosion	23569.3
Well drained + severe erosion	12856.2
Well drained + slight erosion	3894

Hazra *et al.*, 2011

According to the study it has been found that 50.58% of the soil drainage type is well drained and 12.13% of excessively drained. Thus it could be concluded that 62.71% of soil is classified as well – excessively drained soils followed by moderately drained soil drainage of 18.63%. It is also found that the poorly drained soil drainage is 9.23% as well as imperfectly drained 8.90%. It simply means that well drained soils are of highest percentage followed by moderately and excessively drained soil drainage thus not allowing the rainwater to percolate deep into the soil. Study area also witnesses moderate to severe erosion where moderate erosion occurs is 58.78% of area followed by severe erosion of 28.40% and slight erosion is restricted to only 12.31%. Therefore it is inferred from these values that moderate erosion is dominant in Jharkhand followed by severe erosion in most of the areas and only slight erosion in certain areas. Thus it can be inferred that most of the soil in Jharkhand are prone to severe erosion and are also well drained. It can be further inferred that soil type in Jharkhand is not suitable for water recharge. As most of the water goes off as runoff the recharge into ground water is less. Soil in Jharkhand is thus one of the components responsible for water stress in the state. Thus steps should be taken to capture the running water (Hazra *et al.* 2011).

Conclusion

The land capability classification is grouping of land according to its inherent characteristics. It is grouped into eight classes and each class is indicated by numbers from I to VIII. The first four classes of land are suitable for cultivation and the remaining four are not suitable for cultivation but useful for grazing forestry and wildlife.

References

Bhattacharyya, T., Sarkar, D., Sehgal, J., Velaytham, M., Gajbhiye, K. S. 2009. Soil Taxonomic Database of India and the States (1:250,000 scale). NBSS&LUP, 143: 266.

Biard. F. and Baret, F. 1997. Crop Residue Estimation Using Multiband Reflectance. *Remote Sens Environ,* 59: 530-536.

Conservation Measures 2008. A Source Book for Soil and Water Foundation for Ecological Security pp. 46-48.

Dalton, K. L. 1989. A review of information relevant to the riverine woodland and forest rengelands of south-western New South Wales Tech Rep. No. 15 Soil Conservation Service of N.W.S, pp. 305.

FAO, 1976. A framework for land evaluation. Soils Bulletin 32. Food and Agriculture Organisation of the United Nations, Rome, pp 72.

Hazra, M., Avishek, K., Pathak, G. and Nathawat, M. S. 2011. Water Stress Assessment in Jharkhand State Using Soil Data and GIS, *J. Appl. Sci. Environ. Manage.*, 15(1): 63 – 67.

Klingebiel, A. A. and Montgomery, P. H. 1961. Land Capability Classification. Agricultural Handbook No. 210, US Department of Agriculture, Washington, DC, pp. 21.

Laxmanrao S. S. 2001. Land Capability Classification Using Remote Sensing and Geographic Information System. M. Sc. Thesis, Indira Gandhi Agricultural University, Raipur, pp. 44-48.

McBratney, A. B., Field, D. J. and Koch, A. 2013. The dimensions of soil security. *Geoderma* 213: 203–213.

Mongi, H. O. and Huxley, P. A. 1979.. Soils research in agroforestry. ICRAF, Nairobi.

Nair, P. K. R. 1987. Soil productivity under agroforestry. In H. Gholz (ed.). Agroforestry: realities, possibilities and potentials. Dordrecht, Netherlands: Martinus Nijhoff (in press).

Nair, P. K. R. 1984. Soil productivity aspects of agroforestry. ICRAF, Nairobi.

New South Wales Department of Planning, 1988. Rural Land Evaluation.

Panhalkar, S. S., Mali, S. P., and Pawar, C. T. 2012 Morphometric Analysis and Watershed Development Prioritization of Hiranykeshi Basin in Maharashtra, India. *Int J Environ Sci.,* 3: 525-534.

Sharma, H. S. 1972. Land capability classification of the Lower Chambal Valley, Proceeding Symposium on Land Use in Developing Countries, 21st International Geographical Congress, Aligarh.

Siddiqi, N. A. 1971. Land classification for Agricultural Planning - A study in Methodology. *The Geographer*, 18: 25-38.

Tejwani K. G., 1976. Using and Interpreting Soil Information for Land Capability, Irrigability and Range Site Classification and for Highways, (Association of Soil & Water Conservationists & Trainees).

USDA, 2001. Guidelines for soil quality assessment in conservation planning. USDA. Natural Rresources Conservation Sercive, Soil Quality Institute.

USDA. Natural Resources Conservation Service National Soil Survey Handbook, part 622.

Vishwanath, B. and Ukil, A. C. 1943. Soil Map of India. Indian Council of Agricultural Research, New Delhi, India.

Young, A. 1986a. Effects of trees on soils. In R.T. Prinsley and M.J. Swift (eds.), Amelioration of soils by trees: A review of current concepts and practices. London: Commonwealth Science Council.

Young, A. 1986b. The potential of agroforestry as a practical means of sustaining soil fertility. In R. T. Prinsley and M. J. Swift (eds.), Amelioration of soils by trees: A review of current concepts and practices. London: Commonwealth Science Council.

Young, A. 1986c. The potential of agroforestry for soil conservation. I. Erosion control (ICRAF Working Paper No. 42) II. Maintenance of fertility (ICRAF Working Paper 43. ICRAF, Nairobi.

Young, A. 1987. Soil productivity, soil conservation and land evaluation. *Agroforestry Systems*, 5:277-292.

Young, A., Cheatle, R. J. and Muraya, P. 1986. The potential of agroforestry for soil conservation. Part III. Soil Changes under Agroforestry (SCUAF): A predictive model. ICRAF Working Paper 44. ICRAF, Nairobi.

19

Problematic Soils Under Different Agro-ecosystems

Sheela Barla and R. R. Upasani

Introduction

Before start of this chapter some pertinent questions need to be answered. What are different agro ecology and what are the problem soils. Since dawn of agriculture in early 2000 B.C. man has been doing agriculture as subsistence farming to cater the needs for his own consumption. Later on as the population increased, the availability of land per pearson reduced and further the commercialization of agriculture after world war II, particularly in industrialized nation, compelled farmers to increase productivity per unit area. The green revolution involving higher fertilizer uses, more use of agrochemicals together with evolvement of improved crop varieties in later part of twentieth century played vital role in producing higher grain production resulting the global production triplet in contrast to merely increase in total global cultivated area to the extent of 10%.

The catastrophe of green revolution was realized in later years when reports of soil health degradation started appearing. Intensified agriculture characterized by excessive use of fertilizers, agrochemicals and unsupportable level of over exploitation of natural resources led to serious degradation of ecosystem. Thus management of agroecosystem has reached to a crossroad with serious pressure to increase productivity simultaneously holding the soil health with prime concern for future generation. Thus the soil heath under different agroecosystem are posing problem as the soil productivity has not reached to plateau level but has started reducing year after year. Thus such soils are becoming problem which need special attention to bring them back to support higher crop production.

What is Agroecology: Agroecology can be defined as a study of agricultural ecosystem and their components as they function in themselves and as a part of the larger ecosystem. An agro - ecological zone is land unit carved out of agroclimatic zones super imposed on land form which acts as modifier to climate and length of growing period.

1. Cold Arid Eco-region with Shallow Skeletal Soils: The region belongs to north -western Himalayas pertaining to Ladakh and Gilgit districts. The area represents about 4.7 percent of total geographical area of India. The predominant soils of this region are alkaline in nature along with moderate level of organic matter. Soils of the region are quite distinct from those of other zones due to differences in climate, topography, vegetation and rocks. In a study Gupta and Arora (2017) reported that the soils of this region are shallow to very shallow in depth. The colour, texture, structure and permeability varies considerably depending upon topography and vegetation. The soils consist of coarse materials hence are called as skeletal soils mixed with rock pieces. The genesis of free Calcium carbonate is because of weathering of parent materials like limestone, dolomitic limestone and shale. There are three major soil order i.e. Entisols, Inceptisols, Mollisols with occasional presence of Alfisols.

2. Hot Arid Eco-region with Desert and Saline Soils: The region comprises of 9.78 percent of geographical area of India. The major that part falls in this region are western plain, that is south-western parts of Haryana and Punjab, western parts of Rajasthan, the Kachchh peninsula and northern parts of Kathiawar peninsula (Gujarat). The soils of region has sandy soils representing Thar series. They are moderately calcareous and alkaline in nature. The salinity of the soil results in physiological droughts that occur frequently. There is nutrient imbalance of nitrogen, The nutrient imbalance is characterized by phosphorous, zinc and iron in particular.

The major sub group of soils of different regions are as below:

(a) Marusthali: The soils are characterized by shallow and deep sandy desert soils; while the length of growing period is less than 60 days.

(b) Kachchh Peninsula: This region has feature of deep loamy saline and alkali soil, and growing period is also less than 60 days similar to Marusthali.

(c) Rajasthan Bagar, North Gujarat Plain and south-western Punjab Plain: This region is characterized by deep, loamy desert soils which also includes saline phases. The length of growing period ranges from 60 to 90 days.

(d) South Kachchh and North Kathiawar Peninsula: The region has deep loamy saline and alkali soils and the region is characterized by less than 60 days of growing period.

3. Hot Arid Eco-region with Red and Black Soils: The region comprises of part of Deccan plateau covering districts Bellary, south western parts of Bijapur, Raichur and Anantpur. This region covers an area of 4.9 m ha which is nearly 1.5 percent of the total geographical area of India. The length of growing period is more than 90 days. The soils are represented by gently sloping shallow and medium red soils, deep clayey black soil. The dominant red (loamy) soil represented by Kadiri series are slightly acidic and non calcarious. The subdominant clayey black soils represented by Raichur series are slightly alkaline and non calcarious.

4. Hot semi-arid eco-region with alluvium derived soils: This region involves northern plains and central high lands, Gujrat comprising of nearly 32.3 m ha which is about 9.8 percent of total geographical area of India. The soils are deep loamy alluviam – derived soils mostly deep loamy and clayey mixed black soil.

5. Hot semi-arid eco-region with medium and deep black soils: This region comprises of plains of Gujrat, peninsula region of Kathiawar, western Madhya Pradesh, and south eastern part of Gujrat and Rajasthan. The main features of soils of this region is that they are clayey and level to very gently sloping deep black soils. The slightly alkaline calcarious soils with swell and shrink properties of Malwa plateau are main characteristics of soil of this region. The intermittent dry spells are the recurring features of this region. Besides unsupportive drainage restrict roots development and oxygen supply in the roots of plants. Severe alkalinity and sea water inundation is the special characteristics of this region particularly in Kathiawar coastal land area.

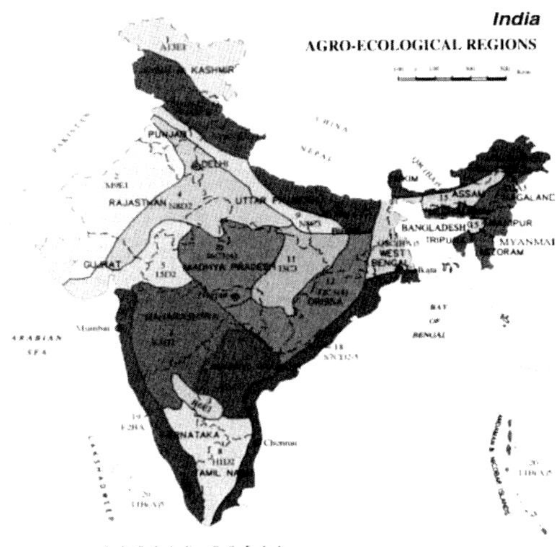

Fig. 19.1: Agroecological regions of India (*See colour plate on page 273*)

6. Hot semi-arid eco-region with shallow and medium (dominant) black soils

The region comprises of Deccan Plateau—most of the central and western parts of Maharashtra, northern parts of Karnataka and western parts of Andhra Pradesh. The area is about 31.0 m ha which is 9.42 percent of total geographical area of India. The major draw back of this situation is that soil erosion results because of heavy run off during stormy cloud bursts. The deficiency of nitrogen, phosphorus and potash poses imbalance in nutrient status of the soils.

7. Hot semi-arid eco-region with red and black soils: The region comprises of nearly 5.2 percent area i.e. 16.5 m ha of total Indian geographical area. The main region under this agro ecological group are Deccan Plateau (Telangana) and parts of Eastern Ghats in Andhra Pradesh. The soils are marginally slopy, and red non calcarious and neutral soils. The black cotton soils are clayey, calcareous and strongly alkaline. The soils with swell and shrink phenomena on wetting and drying (Kasireddipalli series) are main charecteristics of the region. Although they have production potential but reqires heavy management. Soil erosion owing to heavy rum off of soils as well as nutrients are the major draw back of this region. The sub soils salinity and sodicity due to imperfect management of black soils has led to Nitrogen, phosphorous and zinc deficiency, repeated droughts resulting in crop failures occurs.

8. Hot semi-arid eco-region with red loamy soils: The region comprises of Eastern Ghats, southern portions of the Deccan Plateau, Tamil Nadu uplands and western portions of Karnataka. The area is about 19.1 m ha which is about 5.8 percent of total Indian geographical area. The soils are non calcareous representing Tyamagondalu series which are slightly acidic and calcareous soils which are slightly alkaline represented by Palathurai series are existing. Both the soil types are having low cation exchange capacity. The region is characterized by heavy soil erosion due to run off. The crops suffers due to droughts during their growing period as a result of low water holding capacity of soil accompanied by poor soil texture. This also results in soils deficient in nitrogen, phosphorous and zinc causing nutrient imbalance in soils.

9. Hot sub-humid (dry) eco-region with alluvium-derived soils: The region is spread over northern Indo-Gangetic Plains of the western Himalaya which is nearly 3.7 percent of total geographical area of India. The soils are deep , loamy and alluvial type. The soils are represented by Shajadapur, Gurudaspur and Itawa series as well as Basiaram series . They are gently sloppy (Basiaram series) and level Shajadapur, Gurudaspur and Itwa series. The soils have low organic carbon content. The water logging owing to unscientific use of irrigation water are the major drawback of this agro ecological situation.

10. Hot sub-humid eco-region with red and black soils: This agro ecological region occupies nearly 22.3 m ha which is 6.78 percent of total geographical area. The region is spread over Malwa Plateau, Bundelkhand uplands, Narmada valley, Vindhyan scarplands, northern portions of Maharashtra Plateau and some districts of the Madhya Pradesh. The soils are mainly black soils with patches of red soils. The soils of this region are represented by Marha, Kheri and Linga series, with Kamliakheri series which posses slightly alkaline calcarious montmorillonitic soils. They have properties of swell and shrink features, which makes the cultivation difficult. Sometimes dry tillage and interculture in standing crop also makes very difficult to perform. Besides this, crop failure due to water logging during rainy season as well as droughtness during dry period makes the crop production at very high risk. Nutrient imbalances occur due to deficiency of nitrogen, phosphorous and zinc.

11. Hot sub-humid eco-region with red and yellow soils: The region comprises of Chhattisgarh region and southwest highlands of Bihar covering nearly 14.1 m ha which is about 4.3 percent of total geographical area of India. The region is characterized by red and yellow soils which are deep in depth, having calcareous in nature having neutral to slightly acidic in reaction. Water erosion is the main problem of the region. The crop suffers due to water logging during early crop growth while suffers due to water stress during latter part of crop growth. The gravelly like quality of sub soil and coarse texture in some parts restricts water availability to plants. Nutrient imbalances occur due to deficiency of nitrogen, phosphorous, zinc and boron.

12. Hot sub-humid eco-region with red and lateritic soils: The region is spread over an area of 26.8 m ha which is nearly 8.2 percent of total geographical area . The region covers Chotanagpur plateau (Jharkhand), western portions of West Bengal, Dandakaranya and Garhjat hills of the Eastern Ghats of Orissa and Bastar region of Chhattisgarh. The soils are characterized by Pusaro, Bhubaneshwar and Chougel series which are finely clayey in texture, non calcarious, slightly to moderately acidic, having low cation exchange capacity. The depth of soil varies as the ridges and furrows are shallow in depth while in valleys deep soils are found which are main cultivable lands. Severe soil erosion, intermittent drought during growing period of crops accompanied by low water holding capacity of soil owing to gravel nature of soils are the major constraints of this region. The soils are poor in nutrients owing to deficiency of nitrogen, zinc and boron besides fixation of phosphorus is a serious matter of concern.

13. Hot sub-humid (moist) eco-region with alluvium-derived soils: This region occupies an area of 11.1 m ha which is about 3.4 percent of entire geographical area of India. The region occupies northeastern Uttar Pradesh and northern Bihar including Central Himalayan foothills. Soils are characterized as calcareous and moderately alkaline in reaction (Kesarganj and Sabour series). The soils shows different pattern of profile development from A-C soils in the flood plains to A-Bt-C soils on stable terraces. The soils of foot hills of Himalaya (Haldi series) are deep, loamy and rich in organic matter. Limited soil aeration, as a result of recurring flooding, associated with poor drainage are major constraints of the region. Salinity and sodicity affects crop yields. Poor nitrogen, phosphorus and zinc causes nutrients imbalance.

14. Warm sub-humid to humid with inclusion of perhumid eco-region with brown forest and podzolic soils: The region belongs to Jammu and Kashmir, Himachal Pradesh and north western areas of Uttar Pradesh. The area is spread over in 21.2 m ha which is 6.3 percent of entire geographical area of India. The main characteristics of soil of region are that podzolic soils are shallow to deep, medium having high organic matter content and weak (A-C) to well developed (A-Bt-C) horizons (Gogji-Pather Wahthora and Kullu series and others). The soils are fine loamy in texture having neutral pH in reaction and base saturation 50 per cent or more base saturation. The Tarai soils of Nainital and Garhwal districts are deep loamy, neutral to slight alkaline, moderately base saturated soils rich in organic matter. The major problem of the region are that there are limited option for growing crops on higher northern altitudes owing to severe climatic condition as well as poor drainage condition. Deforestation has brought severe soil erosion and degradation of soil. Soil acidity has also developed due to deforestation and excessive slopes specially in Kangra and Manali region. In lower hills particularly in Himachal Pradesh, drought is a common feature owing to excessive run off and coarse soil texture.

15. Hot sub-humid (moist) to humid (inclusion of perhumid) eco-region with alluvium-derived soils: The region is spread over in 12.1 m ha having 3.7 percent area of entire Indian geographical area. The region includes plains of the Brahmaputra and the Ganga river, that is, parts of Assam and West Bengal states. The soils are characterized by slight to strongly acidic and generally have low to moderate base saturation (Jaihing, Kanagarh and Jorhat series). The major problem of this region are flooding and water logging associated with excessive leaching of bases and nutrients has resulted in low base status soils especially in the Brahmaputra plain. Soil acidity is the main cause of nutrient imbalance.

16. Warm sub-humid eco-region with brown and red hill soils: The region is spread over 9.6 m ha, which is about 2.9 percent. The region covers northern hilly parts of West Bengal, northern parts of Assam, Arunachal Pradesh and Sikkim. The major soils occurring in the region are shallow to deep, medium having high organic matter content, and weak (A-C) to well developed (A-Bt-C) horizons. They are classified as Brown Forest and Podzolic Soils. Some such soils (Gogji-Pather, Wahthora and Kullu series) are fine loamy, neutral in reaction and have 50 per cent or more base saturation. Tarai soils (Nainital and Garhwal districts) are deep, loamy, neutral to mildly alkaline moderately base saturated soils high in organic matter and typically represented by Haldi series. The major constraints of this region are characterized by severe climate especially very low temperature in northern high altitude which permits limited option of crops production. Soil erosion due to excessive deforestation and slopy land resulted soil degradation. Such soil degradation often invites land slides, Soil acidity, especially in Kangra and Manali areas of Himachal Pradesh. Drought is experienced especially in the lower hills due to excessive runoff and coarser soil texture.

17. Warm Perhumid eco-region with red and lateritic soils: This agro region comprises of north eastern hills (Purvachal) and the states of Nagaland, Meghalaya, Manipur, Mizoram and the southern parts of Tripura, occupies 10.6 m ha (3.3 per cent of the total geographical area of India). The main feature of soils are that they are shallow to very deep, loamy, red and lateritic and red and yellow soils. Soils of Dialong series are acidic in reaction and have moderate bases on the exchange complex. The major draw backs of the region is characterized by limted choice of crop, owing to steep slope, which encourages excessive run off and hazards due to severe soil erosion. Soil shows poor base saturation as excessive precipitation invites intense leaching. In valleys water stagnation is another problem as post monsoon rain restricts option for crop cultivation. Option for second crop after monsoon is restricted due to very low temperature as a result mono cropping is main practice of this region.

18. Hot sub-humid to semi-arid eco-region with coastal alluvium-derived soils: This region is covered in about 8.5 m ha which is about 2.6 percent of total Indian geographical area. The soils are characterized by Motto and Kalathur series which are slight to moderately sodic and clayey. They are different in their cation exchange capacity. The Kalathur soils have high swell-shrink potential. The draw back of the region are that drainage conditions are not proper as a result poor aeration affect crop production. Such condition also brings about soil salinity and sodicity which affect crop production. The region experiences heavy devastating type cyclones during monsoon periods, as well as returning monsoon.

19. Hot humid perhumid eco-region with red, lateritic and alluvium-derived soils: The region comprises of Sahayadris, western coastal plains of Maharashtra, Karnataka and Kerala and Nilgiri Hills of Tamil Nadu, covering nearly 11.1 m ha which is nearly 3.6 per cent of India's total geographical area.

The main soil features of this eco-region is that they dominate with Sahyadri range and alluviam soils in coastal plains. While Thiruvananthapuram and Kunnamangalam soil series are deep, clayey, strongly to moderately acidic and poor in base saturation. The soils are low in moisture retention capacity as well as poor in soil fertility. The major soil problems are that they are subject to heavy leaching, which results in leaching of nutrients and bases resulting in depletion of nutrients. Similarly water logging also affects poor plant growth in coastal areas. Soil erosion are also one of the major problems in sloppy land due to excessive run off of rain water. This situation creates saline marshy lands in patches.

20. Hot humid per-humid Island eco-region with red loamy and sandy soils: This region involves an area of 0.8 m ha which is about 0.3 percent of entire Indian geographical area. The region comprises of Andaman and Nicobar Islands in the east and Lakshadweep in the west. The soils of Andaman and Nicobar are represented by medium to very deep red loamy soil which also includes marine alluvium-derived soils along the coast. Some soils are also slightly to strongly acidic and moderate to low (40-70 per cent) in base saturation particularly Ahargaon, Dhanikhari and Garucharma series. Highly calcarious soils are found in Lakshadweep Islands. The degradation of rain forests has led to drastic soil erosion, However, introduction of sustainable plantation crops (oilpalm) by using technology may ensure maintaining the ecosystem. Deforestation may be the demand for plantation of oil palms which ultimately protect ecosystem. Submergence of coastal lands develops saline marshy lands resulting in formation of sulphates soils.

Management of problematic soils

The objective of a sound management programme is to increase and sustain efficiency in the production of agricultural crops. Until 1900 a large part of the increase in agricultural production was affected by bringing large area and more land under cultivation. Very few unused land remains today in many developed countries, and it appears that future increases in production will come about largely by improvement in crop yields. Consequently, raising yields by better fertilizer and management practices is of paramount importance.

The 'problem soil' herein means the soil that has agricultural problems due to the soil's unsuitable physical and chemical properties, or less suitable for cultivation, resulting that crops are not able to grow and produce yields as normal. These soils always occur naturally, including saline soil, acid sulfate soil, sandy soil, organic soil, skeletal soil and shallow soil. Furthermore, it may also include areas with steep slope. If these lands are used for agricultural purpose, then it may cause some severe effects on the ecology and environment.

The problem arises from the acid sulphate soils

The most important characteristics are a field pH of below 4, owing to the oxidation of pyrite to sulphuric acid, and a generally high clay content. If samples of the pyrite layers are air-dried in the laboratory, the pH may drop by a further 2 units. Other properties such as organic matter content and cation exchange capacity may vary widely.

(*See colour plate on page 273*)

The strong acidity of an acid soil affects the availability of various nutrients such as nitrogen, phosphorus, potassium, sulfur, calcium and magnesium to the plants, resulting in the shortage of these elements in plants, so they cannot grow normally. In strongly acid soil, iron and aluminum may dissolve in the soil to the levels that are toxic to many crops as well as soil microorganisms. Water in an acid sulfate soil area is normally astringent (sour) and unsuitable for agriculture and consumption. In a fish pond, there might be the toxicity of hydrogen sulfide gas, carbon dioxide and organic acids.

Soils are neither "good" nor "bad" because the distinction is often based on their intended use. However many soils have characteristics that make specific management interventions desirable to avoid problems for agricultural production or environmental degradation. An arbitrary differentiation is made between "**problem soils**" in which the soil characteristics themselves pose problems for their optimal use, and "**degraded soils**" in which unwise management interventions create supplementary environmental and productive problems.

Management of Highly Organic Soils

Soils rich in organic matter and un-decomposed plant material are Histosols. They occur particularly in areas where decomposition of organic matter is hampered by cold temperatures (in Boreal climates) or where the decomposition is hampered by continuous wet conditions (in the Wet Tropics). The fertility of Histosols is normally low when the natural vegetation is abruptly replaced with agricultural crops. The cycling of plant nutrients is interrupted and leads to chemical exhaustion. Especially micronutrients such as boron, copper and zinc may be lacking.

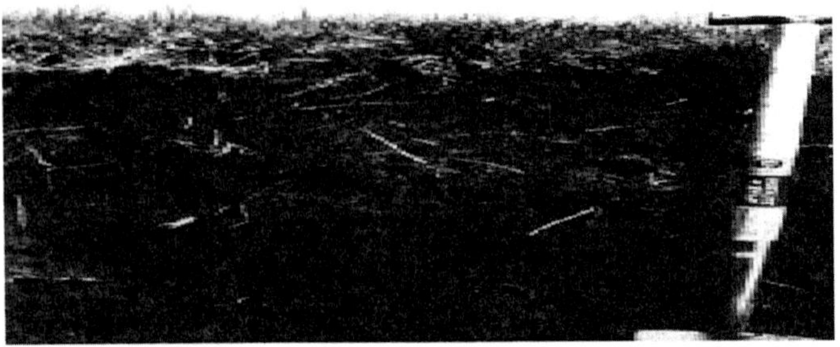

(*See colour plate on page 273*)

To reclaim Histosols shallow drainage ditches need to be built. The natural vegetation is left standing for a while in order to speed up the drying of the peat. Drains can be 1 metre-deep at 20-40 metre intervals. Construction of a complex drainage system at the start of reclamation should be avoided because this may cause uneven subsidence of the land, which will disrupt the connections between sucker drains and collecting drains. Small-scale farmers sometimes carry out controlled burning of the peat to free nutrients and to raise the pH of the surface soil. Burning stimulates plant growth but burning and its precise effects are still open to discussion.

Management of heavy cracking clays

Soil water management, tillage, cropping systems, and nutrient management pose special problems in heavy cracking clays. These soils mainly belong to Vertisols_and vertic subgroups of other soils and occur mainly in (sub) tropical areas with a pronounced dry season. Management of soil water is the important aspect of soil management in the semi-arid tropics. The poor internal drainage and extremely slow hydraulic conductivity, leads to water logging, and delay in planting. Extreme consistency properties of the soil permit tillage operations within a narrow soil moisture range only: because soils are sticky when wet and hard when dry. Use of tillage implements under wet conditions may result in soil sticking to implements and the formation of large clods. Wide deep cracks in the dry season, permit easy entry of rainfall and water moves freely into the cracks. Although chemically rich these soils under sustained, high-input systems, may suffer from fertility problems on account of limited availability of N, P and micronutrients. The presence of heavy clays may be associated with a micro relief called gilgai that is a result of the continuous churning of these soils. Infrastructure and buildings may seriously be damaged in the long run, if they are built on these soils. Besides the need to get as much of the rain as possible into the soil for use by the crop, there is the need to provide adequate surface drainage to avoid plant injury or slow growth from water logging once the cracks have closed and infiltration rates have slowed down. A traditional and relatively early method was the cambered bed. A cambered bed can be formed by ploughing up and down so that the soil is turned inwards to the centre. Cambered beds have been used successfully in many areas of Africa. In very dry areas, water logging is less likely to be a problem, and tillage is important to get every drop of water in the soil, and minimize runoff and evaporative losses. The roles of ridges and furrows are reversed: water is designed to run off the ridge, sometimes suitably broadened, which then becomes a water harvesting device designed to lead runoff into the furrow in which the crop is planted. To block the water movement in the furrow, the ridges may be 'tied' at intervals with a cross ridge, although in occasional unusually wet years the ridges may be untied.

(*See colour plate on page 274*)

Cropping system is an important aspect of heavy cracking clays management. The ICRISAT system of post-harvest ploughing followed later by seedbed preparation and dry seeding before the rains has been found to be successful in regions with predictable rains. Proper management and timing of cultivation practices are crucial for the management of heavy cracking clays.

Management of gypsiferous soils

Soils with more than 25 percent of gypsum may hamper plant growth. The soil material then lacks plasticity, does not stick together and becomes completely unstable in water. Consequently erosion of gypsiferous soils can be very severe. Soils with significant amounts of gypsum particularly occur in the driest areas on earth.

(*See colour plate on page 274*)

Improvement in the productivity of gypsiferous soils in rainfed conditions may be achieved by:

- Soil terracing on the deep hilly soils to prevent erosion.
- Supplementary irrigation where water resources are available.
- Harrowing the land after harvesting and before the rainy season to improve the infiltration of water and conserve soil moisture.
- Replacement of fallow by small-grain leguminous crops in wheat-fallow rotations to improve soil organic matter.
- Subsoiling to break the cemented gypsic subsoil to improve root penetration and reduce susceptibility to drought.
- Use of fertilizers, especially nitrogen and phosphorus for cereals.

Where irrigation water is available, leaching of gypsum is necessary to keep the salt content low. An effective drainage system is required to maintain a relatively low water-table and keep salinity under control. Cavities created by leaching gypsum from the surface soil make it necessary to level the land surface each year. Irrigation canals must be lined to prevent the canal walls from caving in. Soils with a cemented gypsic layer may impede the installation of drainage systems. Under irrigated agriculture on gypsiferous soils with a low level of organic matter and total nitrogen, the regular application of N fertilizers is essential to secure adequate yields of most crops. The potential productivity of gypsiferous soils is related to the depth of the gypsic layer. In soils with a gypsic layer below 60 cm depth, the plant roots penetrate freely and there are sufficient soil volume for nutrients. Fertilization of these soils improves plant growth and increases yield. In shallow soils, with a gypsic layer near the surface, the soil volume is limited and plants do not thrive generally.

When Gypsiferous soils contain only little gypsum in the upper 30 cm soil layer, they can be used for production of small grains, cotton, alfalfa, etc. Dry farming on deep gypsiferous soil requires use of fallow years and water harvesting techniques, but is rarely rewarding under adverse climate conditions. Many gypsiferous soils in (young) alluvial and colluvial deposits have relatively little gypsum. Such soils can be very productive if carefully irrigated. Even soils containing 25 percent powdery gypsum or more may still produce excellent yields of alfalfa hay, wheat, apricots, dates, maize and grapes, if irrigated at high rates in combination with forced drainage. Application of fertilizers is required for good yields. Most areas of gypsiferous soils are in use for extensive grazing.

Management of calcareous soils

(*See colour plate on page 274*)

Calcareous soils have often more than 15% $CaCO_3$ in the soil that may occur in various forms (powdery, nodules, crusts etc…). Soils with high $CaCO_3$ belong to the Calcisols and related calcic subgroups of other soils. They are relatively widespread in the drier areas of the earth. The potential productivity of calcareous soils is high where adequate water and nutrients can be supplied. The high calcium saturation tends to keep the calcareous soils in well aggregated form and good physical condition. However where soils contain an impermeable hard pan (petricalcic horizon) they should be deeply ploughed in order to break the pan. This should be followed by the establishment of an efficient drainage system. Furrow irrigation is better than basin irrigation on slaking calcareous soils. On undulating lands, contour and sprinkler irrigations are better options than flood irrigation. Drip irrigation may also be practiced. Calcareous soils generally have low organic matter content and lack nitrogen. Nitrogen fertilizer may be applied any time from just before planting up to the time the plant is well established. Application of nitrogen through side-dressing to the growing crop is an efficient way of nitrogen application. Care should be exercised so as not to apply nitrogen close to the seed as it may prevent germination. Ammoniac sources of nitrogen and urea should not be left on the surface of calcareous soils, since considerable loss of ammonia through volatilization may occur, and they should be incorporated in the soil instead. Phosphorous is often lacking in calcareous soils. Amounts to apply depend on how deficient the soil is and the crop requirements. Excess applied phosphorus may lead to deficiency of zinc or iron. To be effective on calcareous soils, applied phosphorus fertilizer should be in water soluble form. Band application of phosphate is more effective as compared to broadcast application. Application at the time of seeding has been found to be most appropriate since phosphorus is required mostly during the younger stages of plant growth.

Calcareous soils usually suffer from lack of micronutrients, especially zinc and iron. Zinc deficiency is most pronounced in maize, especially under high yield intensive cultivation systems. Zinc sulphate is an effective zinc source and is the most popular form in use. For soil application, zinc sulphate is broadcasted and incorporated in soil. A single application lasts for several years. Foliar applications of zinc are used on fruit trees. Heavy applications of animal manure are helpful in preventing deficiency of iron and zinc.

Acid soils

Acrisols Alisols Podzols

(*See colour plate on page 274*)

Acid soils are those that have a pH value of less than 5.5 for most of the year. They are associated with a number of toxicities (Aluminum) as well as deficiencies (Molybdenum) and other plant restricting conditions. Many of the acid soils belong to Acrisols, Alisols, Podzols and Dystric subgroups of other soils. An extreme case of an acid soil is the acid sulphate soil (Thionic Fluvisols and Thionic Cambisols).

There are two main belts of acid soils:

- In the humid northern temperate zone, which is covered mainly by coniferous forests; and
- In the humid tropics, which is covered by savannah and tropical rain forest.

Acid sulphate soils are usually left under natural vegetation or used for mangrove forestry. If water is managed well they can support oil palm and rice. Some other crops grown on acid soils around the world include: rice, cassava, mango, cashew, citrus, pineapple, cowpeas, blueberries and certain grasses.

An integrated approach to acid soil management comprises a spatially variable liming strategy, the use of acid-tolerant species, efficient use of fertilizers, suitable crop rotations and crop diversification. Soil testing needs to be carried out every two to three years to determine the lime requirements of the field. The buffering capacity needs to be assessed to work out the amount of lime needed to neutralize soil acidity to the desired level. The negative effects of soil acidity on physical and chemical soil conditions can be partly compensated by ensuring high organic matter content.

Acid sulphate soil management is more delicate and has to be based on cautious water management in order to prevent oxidization processes of pyrite:

- The first strategy is to drain and completely oxidize the soil, and then flush the acidity formed out of the soil. This strategy solves the problem once and for all but has severe disadvantages: it is expensive, poses a threat to the environment (acid drain water!) and depletes the soil of useful elements together with the undesirable ones. Liming of drainage water has been applied to reclaim Acid Sulfate soils in Australia.
- The second strategy is to try to limit pyrite oxidation by maintaining a high groundwater table. A precondition is the availability of sufficient water. This method also requires substantial investments in water management, while the potential danger of acidification remains present. This strategy is widely followed, both in temperate regions and in the tropics, often with ingenious adaptations to suit local conditions and practices.

Incorporation of lime or dolomite into the upper cultivable soil layer is an effective method for amelioration of acid soils. Banding or pelleting lime onto the seed at sowing is also a common practice used to aid with the establishment of temperate pasture legumes. Lime can also be applied as a preventative treatment for soil infertility, and to supply calcium and magnesium to deficient soils. Liming raises the pH of acid soil, thus the action of nitrogen fixing bacteria becomes uninhibited and nitrogen fixation increases. Nitrogen mineralization from plant residues and organic matter has been reported to increases when lime is applied to acid soil. Although lime is primarily applied to raise soil pH and amend toxicities associated with acid soil, liming has also been used to improve soil structure.

Management of Sandy Soils

Sandy soils are those that are generally coarse textured until 50 cm depth and consequently retain few nutrients and have a low water holding capacity.

Soil management practices which lead to an increase in the fine fraction are helpful in improving soil properties and crop productivity.

Fertilization of these soils is considered essential. Inorganic fertilization is the main practice.

(*See colour plate on page 275*)

- Application of organic manures can supply nutrients in slowly available forms and improve soil physico–chemical properties.
- Surface application of organic manures to sandy soils does not last so the manure should be dug deeper into the soil or a carpet-like layer spread of not less than one centimeter thick, which will improve water storage, biological activity, nutrient status and increase yields.
- Mulch can be added to improve water storage by reducing evaporation. Crop residues on the surface of the soil reduce evaporation losses, decrease the range between maximum and minimum soil temperature, and reduce wind erosion.
- For tillage to be really effective, it has to be done at the earliest possible time after irrigation or rainfall when the evaporation rate is still high.
- Minimum tillage, maintenance of a cover crop, strip cropping, crop rotations, control of grazing and establishment of shelter belts and windbreaks are some of the protective measures to counter the high susceptibility of sandy soils to erosion.
- Besides the conventional dry vegetation method, use of artificial surface sealants such as petroleum, synthetic rubber, chemicals and water soluble plastics have also been adopted for dune and drift sand stabilization.
- Afforestation with selected trees and shrubs is a complementary measure that should follow stabilization of dunes.

- Overgrazing on coarse textured soils must be avoided. The introduction of rotational grazing helps to combat this hazard. It might be better not to permit grazing but to use fodder cut on feeding lots.

Management of Salt Affected Soils

When salts more soluble than calcium carbonate and gypsum are present in the soil and affect crop growth and yield of most crops, these soils are considered salt affected. Most of these soils have an Electrical Conductivity of more than 4 Ohms/cm. Many of them are classified as Solonchakz. The presence of salts affects the plant uptake of nutrients and the microbiological activity in the soil. Salinity may also affect other soils to a lesser extent and may; lead to recognition of saline phases which also deserve attention when present under salt-sensitive crops (spinach, etc...).

(*See colour plate on page 275*)

Methods adopted to remove excess salts from the soil surface and the root zone in saline soils includes:

To prevent the excessive accumulation of salt in the root zone, irrigation water (or rainfall) must be applied in excess of that needed for the evaporation of the crop. Leaching can be timed to precede the critical growth stages at which stress should be prevented. This can be timed through irrigation during dry seasons. Leaching at times of low evapo-transpiration demands is more efficient, for example, at night, during high humidity, in cooler weather or outside the cropping season. Leaching is only effective when salty drainage water is discharged through subsurface drains that carry the leached salts out of the area under reclamation (But one should avoid to contaminate other areas under cultivation downstream).

Conclusion

It is now well established that soil is one of the prominent resources for agricultural production. Study of different agroecological situations suggests ways and means to prepare a successful crop planning based on local soil and climatic condition with special consideration of major hurdles with view to rectify them. This should be strategy for exploiting renewable resources on which our nation must built and grow to fulfill all the cherished dreams.

Colour Plates

Chapter 3: Area of Distribution of Wasteland and Problem Soils in Jharkhand

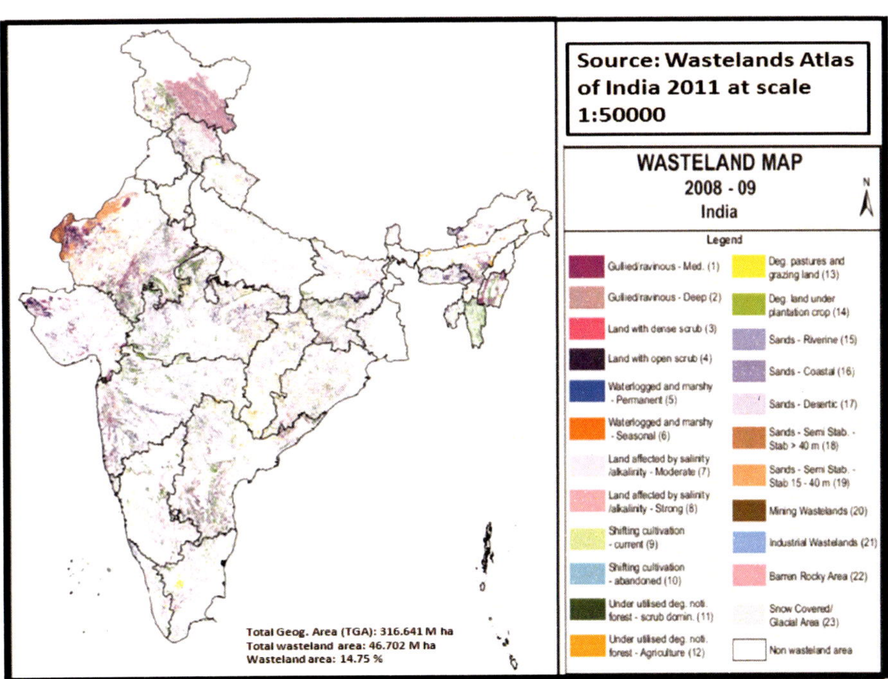

Fig. 3.1: Wastelands Map of India *(For black and white version, please visit page 28)*

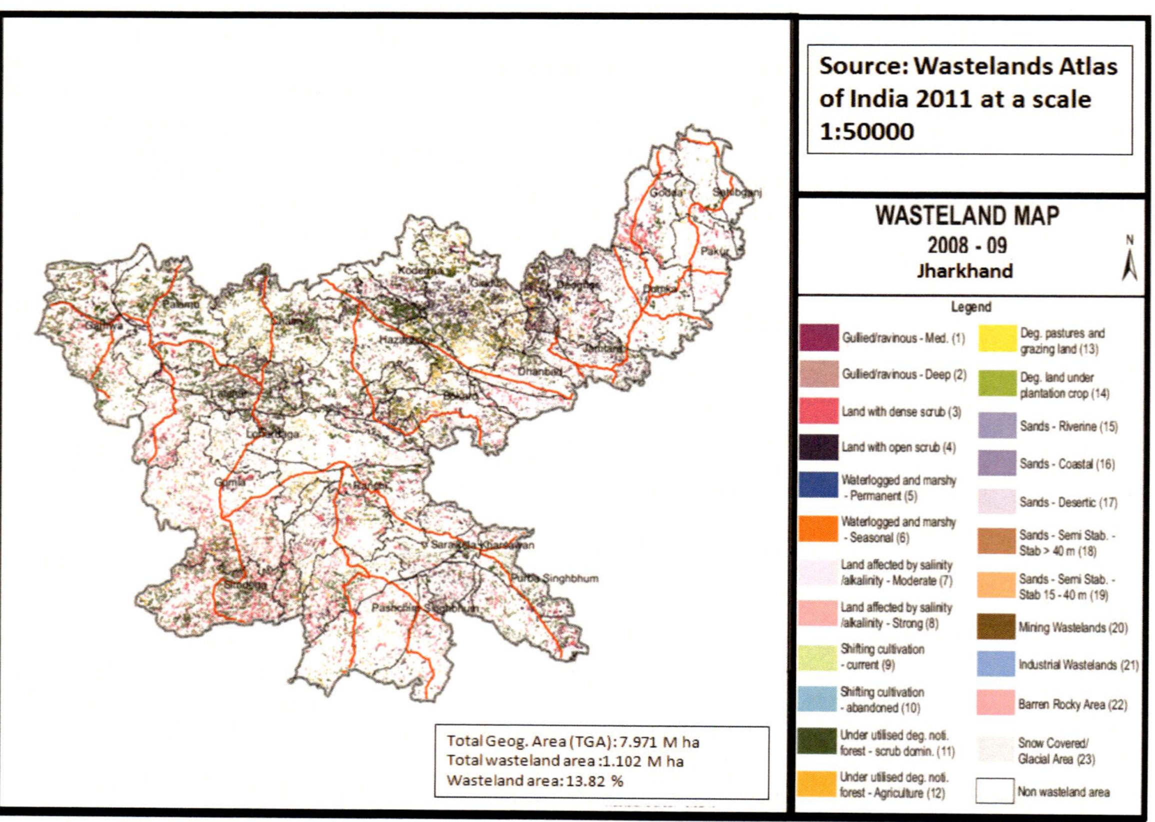

Fig. 3.2: Distribution of Wastelands in Jharkhand *(For black and white version, please visit page 32)*

Chapter 19: Problematic Soils Under Different Agro-ecosystems

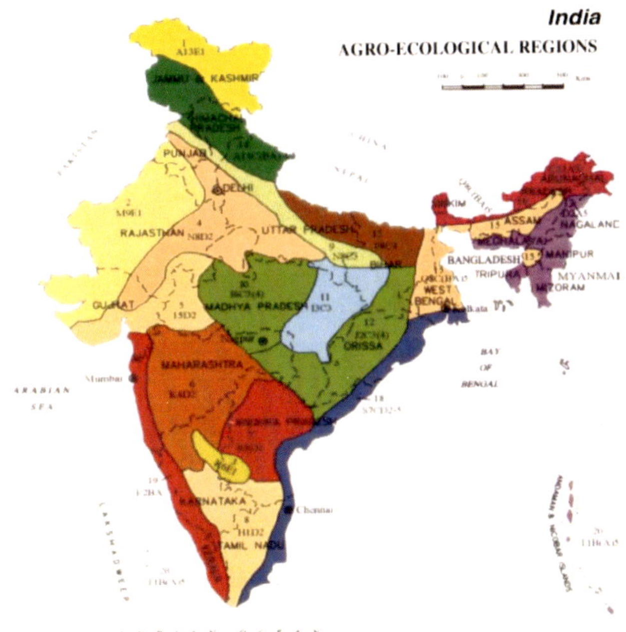

For black and white version, please visit page 255)

(For black and white version, please visit page 261)

(For black and white version, please visit page 262)

(For black and white version, please visit page 263)

(For black and white version, please visit page 264)

(For black and white version, please visit page 265)

Acrisols **Alisols** **Podzols**

(For black and white version, please visit page 266)

(For black and white version, please visit page 268)

(For black and white version, please visit page 269)